T5-CFQ-588

SOCIETY FOR EXPERIMENTAL BIOLOGY

SEMINAR SERIES · 2

EFFECTS OF POLLUTANTS ON AQUATIC ORGANISMS

EFFECTS OF POLLUTANTS ON AQUATIC ORGANISMS

Edited by

A. P. M. LOCKWOOD

Reader in Biological Oceanography, University of Southampton

CAMBRIDGE UNIVERSITY PRESS

CAMBRIDGE

LONDON · NEW YORK · MELBOURNE

Published by the Syndics of the Cambridge University Press
The Pitt Building, Trumpington Street, Cambridge CB2 1RP
Bentley House, 200 Euston Road, London NW1 2DB
32 East 57th Street, New York, NY 10022, USA
296 Beaconsfield Parade, Middle Park, Melbourne 3206, Australia

First published 1976

Printed in Great Britain at the
University Printing House, Cambridge
(Euan Phillips, University Printer)

Library of Congress Cataloguing in Publication Data

Main entry under title:

Effects of pollutants on aquatic organisms.

(Society for Experimental Biology seminar series; 2)

Lectures presented at a seminar held Apr. 11, 1975, at Liverpool University.

Includes bibliographical references and index.

1. Aquatic ecology – Addresses, essays, lectures.
2. Pollution – Environmental aspects – Addresses, essays, lectures. I. Lockwood, Antony Peter Murray.
II. Series: Society for Experimental Biology (Gt. Brit.).
Society for Experimental Biology seminar series; 2.
QH541.5.W3E35 574.5′263 75-32448

ISBN 0 521 21103 4 hard covers
ISBN 0 521 29044 9 paperback

CONTENTS

CONTRIBUTORS

Addison, R. F., Fisheries and Marine Service Marine Ecology Laboratory, Bedford Institute of Oceanography, Dartmouth, Nova Scotia B2Y 4A2, Canada.

Ahern, T. P., Fisheries and Marine Service Marine Ecology Laboratory, Bedford Institute of Oceanography, Dartmouth, Nova Scotia B2Y 4A2, Canada.

Bengtsson, B-E., National Swedish Environment Protection Board, Brackish Water Toxicology Laboratory, Studsvik, S-611 01 Nyköping, Sweden.

Bryan, G. W., Marine Biological Association, The Laboratory, Citadel Hill, Plymouth PL1 2PB, UK.

Corner, E. D. S., Marine Biological Association, The Laboratory, Citadel Hill, Plymouth PL1 2PB, UK.

Darcy, K. The Rosenstiel School of Marine and Atmospheric Science, University of Miami, 4600 Rickenbacker Causeway, Miami, Florida 33149, USA.

Gibson, V. R., Woods Hole Oceanographic Institution, Woods Hole, Massachusetts 02543, USA.

Grice, G. D., Woods Hole Oceanographic Institution, Woods Hole, Massachusetts 02543, USA.

Harris, R. P., Marine Biological Association, The Laboratory, Citadel Hill, Plymouth PL1 2PB, UK.

Howells, G., Central Electricity Research Laboratories, Leatherhead, Surrey KT22 7SE, UK.

Hughes, G. M., Research Unit for Comparative Animal Respiration, University of Bristol, Woodland Road, Bristol BS8 1UG, UK.

Ikeda, T., The Rosenstiel School of Marine and Atmospheric Science, University of Miami, 4600 Rickenbacker Causeway, Miami, Florida 33149, USA.

Larsson, Å., National Swedish Environment Protection Board, Brackish Water Toxicology Laboratory, Studsvik, S-611 01 Nyköping, Sweden.

Lloyd, R., Ministry of Agriculture, Fisheries and Food, Salmon and Freshwater Fisheries Laboratory, 10 Whitehall Place, London SW1A 2HH, UK.

Mackie, P. R., Ministry of Agriculture, Fisheries and Food, Torry Research Station, Aberdeen AB9 8DG, UK.

Reeve, M. R., The Rosenstiel School of Marine and Atmospheric Science, University of Miami, 4600 Rickenbacker Causeway, Miami, Florida 33149, USA.

Svanberg, O., National Swedish Environment Protection Board, Brackish Water Toxicology Laboratory, Studsvik, S-611 01 Nyköping, Sweden.

Swift, D. J., Ministry of Agriculture, Fisheries and Food, Salmon and Freshwater Fisheries Laboratory, 10 Whitehall Place, London SW1A 2HH, UK.

Vandermeulen, J. H., Fisheries and Marine Service Marine Ecology Laboratory, Bedford Institute of Oceanography, Dartmouth, Nova Scotia B2Y 4A2, Canada.

Walter, M. A., The Rosenstiel School of Marine and Atmospheric Science, University of Miami, 4600 Rickenbacker Causeway, Miami, Florida 33149, USA.

Whittle, K. J., Ministry of Agriculture, Fisheries and Food, Torry Research Station, Aberdeen AB9 8DG, UK.

PREFACE

Most fields of scientific endeavour change in direction and technical approach during the course of time. Such a shift in emphasis is currently apparent in pollution studies, the earlier overwhelming concentration on lethal levels and degree of accumulation of toxic materials by organisms now being matched by experimental investigations into the more subtle influences of potential pollutants.

On 11 April 1975 a seminar was arranged at Liverpool University under the auspices of the Society for Experimental Biology, at which a number of the leading exponents of the experimental approach to pollution forgathered to outline the current state of the art. Their lectures, including material on such wide-ranging aspects as the pathways and rates of uptake, detoxification and release from the body of foreign substances, effects of toxins on specific physiological systems and adaptation of organisms to chronic low levels of pollutants, are now assembled in this volume. Coverage is such as to include mention of most of the substances currently considered to have significant effects on aquatic organisms, and the authors have in general included some background review material to put their contribution in perspective. Whilst it is impracticable to give a comprehensive account in a book of this size, the variety and breadth of approach will hopefully provide an introduction to this rapidly expanding field suitable for undergraduates and research students. The numerous and obvious gaps in knowledge made apparent may in addition serve to stimulate other scientists to interest themselves in this area.

Advice from many sources was received during the planning and execution of the seminar and, in addition to the speakers, I would like to thank Professor R. Fänge, Dr G. Howells, Dr J. S. Alabaster and Dr I. C. White for suggestions on contributors; the SEB Zoological Secretary Dr P. Spencer Davies for administrative assistance and Dr T. A. Mansfield for his efforts in successfully garnering the funds necessary to mount the seminar.

The Society for Experimental Biology would like to acknowledge, with gratitude, financial assistance from the following sources towards the cost of running the seminar: Technicon Ltd, Varian Associates Ltd, C. F. Casella & Co. Ltd, T.E.M. Sales Ltd, the British Council, Gelman Hawksley Ltd, Central Electricity Generating Board, the Agricultural Research Council.

July, 1975

A. P. M. Lockwood
Editor for the Society for Experimental Biology

GWYNETH HOWELLS

Introduction

Introducing this volume, which focuses on experimental studies in the aquatic environment, provides me with an opportunity to raise some issues that may perhaps stimulate discussion. Increasing concern for environment and pollution has been influential in changing attitudes to pollution in UK as in other countries, and this is demonstrated by provision in legislation recently enacted, which will strengthen official demands for monitoring and surveillance schemes designed to control unwelcome emissions.[1] Further, as a partner in the European Economic Community we are being pressed to accept quality standards – at least initially for drinking water and foods and perhaps, at a later stage, environmental quality standards. It is also true to say that the improving conditions (e.g. of air and river quality in London) induce further popular expectations of improving quality which are hard to deny, even when economic cost is high.

It is worth reiterating at this stage the sequential steps involved in recognising pollution, setting up relevant investigations and instituting control or management procedures. These are set out in Fig. 1.

In the first place, recognition of 'pollution' surely depends upon the observation of environmental or biological damage and this leads us to identify the toxic agent or contributing circumstances. We can then begin to make measurements of the damaging and background concentrations to which organisms are exposed. At this stage surveillance or monitoring programmes and control measures are sometimes set up, although without an understanding of dose/response relationships, of mechanisms of toxicity, or assessment of the overall risks, these can only be accepted if justified by the gravity of the circumstances.

I now wish to examine in more detail some of the other contributions to pollution studies that are set out in Fig. 1.

Many problems come to mind: How can monitoring schemes be designed? Should they simply be instruments of control (is the emission acceptable; below a set concentration and volume, or not)? How will toxic concentrations be related to effects? How will ecological effects be measured? How will

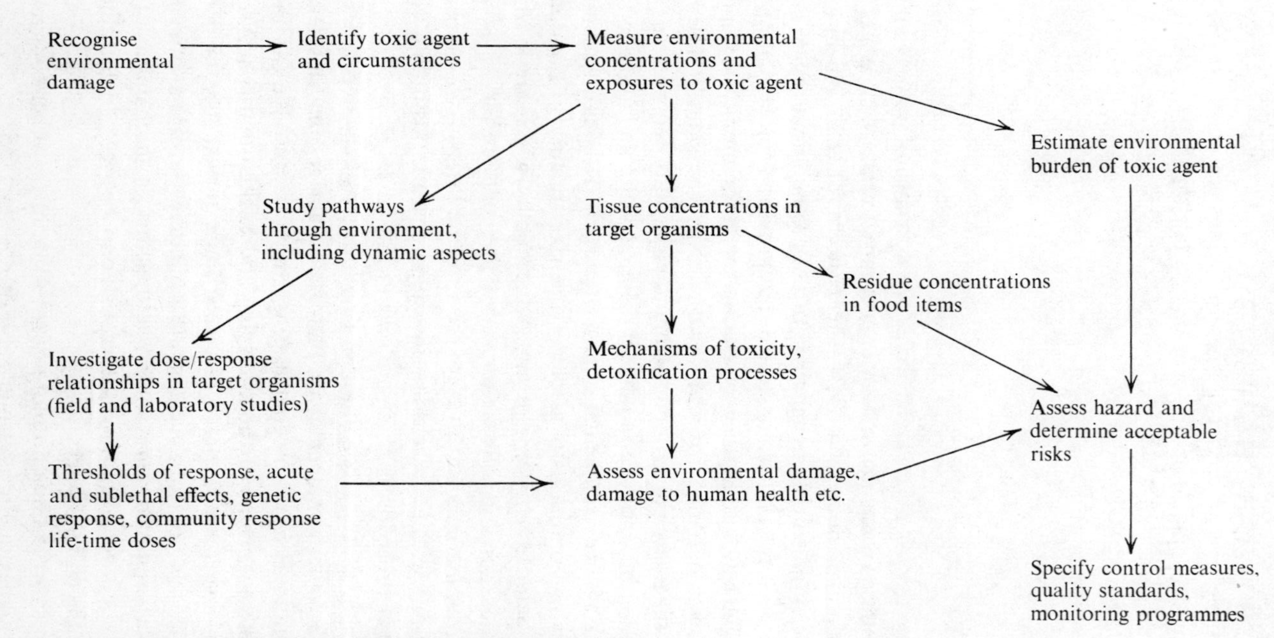

Fig. 1. Scheme of pollution investigation and assessment.

especially sensitive species or populations be protected? What are the criteria of 'acceptable' risk or damage? These questions are not simply a matter for scientific endeavour, common-sense or administrative judgement – they call for an amalgam of all these, and probably many other, approaches.

Once we have observed an effect and related it to a particular toxic agent, it is customary and practical to use sensitive physical or chemical analysis to determine the levels of concentration. However, danger lies in the increasing sophistication and sensitivity of these techniques which engender the view that accurate measurements and an impressive array of decimal places are to be associated with significance. A paramount need is to relate these measurements (concentrations) to their effects in a quantifiable and predictable way, including measurement of length and mode of exposure (dose), so that an assessment can be made of whether a particular observed level is one to elicit remedial action, preventive measures or other control sanctions. A difficulty is that while some pollutants have no natural function (e.g. the man-made element plutonium) many others are present naturally, or even have an essential role and are toxic only because they are present in the 'wrong' form, place or concentration, so that simple detection of their elemental presence may be irrelevant.

Dose/response relationships call, in the first instance, for detailed laboratory studies involving different life stages of organisms and different and varying external conditions and analysis of synergistic or antagonistic reactions. We know a little about the dose/response relationships of only relatively few toxic agents and target organisms and, with present and foreseeable scientific effort, only limited advances can be made towards the ideal state of knowledge.

Moreover, pollutants are seldom presented in real life in the way in which we usually design experiments, i.e. with exposure at a constant concentration. The common pattern of pollutant exposure in the field is to fluctuating rather than constant concentrations, allowing opportunity for recovery, adaptation or simple avoidance of stress, provided the exposure concentration is not acutely toxic. In the case of low subacute exposure, assessment of the integrated effect over the life span is needed, not just an averaged exposure applied for the whole life span.

Responses of individuals (usually the concern of the experimenter and the toxicologist) need to be distinguished from community responses (the concern of the ecologist and the epidemiologist). In the first case, 'averaged' responses are usually irrelevant in the assessment of individual risk, although a 'norm' against which to judge response is a valuable reference standard. Further, although some biological phenomena observed in a group or population can be represented by a normal or log-normal distribution of points, as many examples of non-Gaussian distribution could be cited. Lack of an observed response could thus be because it is below the levels of detection or recognition, because there is a threshold of response, or because the sensitivity of the target organism is anomalous.

Indeed the uncharacteristic sensitivity of a few individuals has the most important implications for species survival and examples of 'naturally' selected tolerant varieties (e.g. sulphur dioxide tolerant Helmshore ryegrass, copper tolerant *Ectocarpus*) are recognised, and in some instances can be genetically distinguished.

Community response assessed in terms of averaged exposure is easier to accept, in that it is the overall status or health of the community that is important. Nevertheless, the survival of a few, more tolerant individuals in a community has provided the means of population recovery when accidents have seemed, at first sight, to have brought about total destruction.

Clearly there are so many complex issues that some rationalisation and selection has to be made if scientific effort is to be focused on major problems. It has been suggested [2] that of eighteen identified 'environmental pollutants' only eight could be considered as potentially hazardous (to man); a critical reappraisal of candidate toxic agents could help to establish a priority listing so that scientific effort could be directed to the most pressing problems. Too often the demand for control of a pollutant appears to be determined by political feasibility, expedience, emotions of the moment, and on unsupported speculation rather than on hard scientific facts. This may cause some suspected agents to be controlled unnecessarily rigorously, while others may be equally unjustifiably exonerated. There is no real substitute for acquiring the relevant information, at whatever level of detail seems appropriate, from both experimental and field observations. Furthermore, if the estimated degree of damage is required to implement control (e.g. if it is to be used in a legal sense to assess 'damages'[3]), or to estimate potential for recovery, some quantitative measure of effect has to be made.

The prospect of studying a wide variety of materials, at a range of concentrations, in conjunction with all environmental variables as well as other pollutants, on all life stages of all possible target organisms, in both acute and subacute dosages, is clearly impracticable. No nation could justify putting so much scientific effort to such an endless task, and ways to limit endeavour to crucial aspects have to be sought. It is salutary to consider the disciplined approach made by toxicologists with man as the prime target, and by radiation health physicists to analyse critical pathways.

Fortunately, we have recourse to some kinds of investigation which although presenting problems of their own, and equivocal if left to stand alone, can help to evaluate hazards and to relate laboratory studies to environmental conditions. First, field studies can be made of existing discharges and the biological observations related to measured gradients of exposure concentration. These investigations have the advantage that contemporaneous conditions and interactions are taken into account, but difficulties of interpretation arise if exposure fluctuates, even when the fluctuations can be recorded. Secondly, investigation of accidents or incidents when excessive quantities of a toxic agent have been released can be revealing not only about the effects of a sudden acute and high-level exposure (and its long-term consequences), but also about the course and time of recovery. Case histories of these occasions are a valuable source of information. Thirdly, continuing surveillance of biological 'status' coupled with water quality or similar type information, can be used to document long-term trends, even where levels of exposure are too low for direct observation of acute effects. These studies, in particular, take account of interspecies interactions, life history and lifetime effects, and can be used to signal unsuspected hazards. Real problems are recognised, however, in selecting valid methods of sampling where discontinuous or sparse populations occur, in determining the level of species identification needed (genus, species, variety) and in data handling (diversity indices)[4]. A parallel may be drawn with epidemiological studies, which although difficult to set up and maintain, can nevertheless provide valid conclusions and risk assessments.

The most promising path lies surely in developing a fruitful interaction between laboratory and field investigations – field observations such as those described, leading us to ask questions that can be answered, albeit only partially, in the laboratory. We look to experimentation to simulate or enhance environmental conditions while excluding or controlling other contributory factors, or to manipulate field conditions, as well as to record and quantify responses. In turn, results of experimental studies should lead to hypothesis and prediction about the environment which can only be confirmed or rejected by further field study.

Most scientists, in the course of their careers, develop interests directed towards either the field or the experimental approach and we should be cautious of conclusions about the living world derived from a single aspect of scientific endeavour. I hope that the initiative shown by the Society for Experimental Biology in promoting this Seminar will help to generate the reaction between these two kinds of investigation.

References

[1] *Monitoring the Environment in the U.K.* Department of the Environment, London, 1974.
Water Act 1973, Chapter 37. HMSO, London, 1973.
Control of Pollution Act 1974, Chapter 40. HMSO, London, 1974.
Three Issues in Industrial Pollution. Second Report of the Royal Commission on Environmental Pollution, 1972. HMSO Cmnd. 4894, 1972.

[2] Stokinger, H. E. 'Sanity in research and evaluation of environmental health'. *Science*, **174**, (1971) 662–5.

[3] *Pollution in some British estuaries and coastal waters. Third Report of the Royal Commission on Environmental Pollution, 1972.*
HMSO Cmnd. 5054 (Minority Report p. 74), 1972.

[4] Orians, G. H., Margalef, R., May, R. M., Whittaker, R. H., Kerr, S. R. and Goodman, D. (contributors).
'Diversity, stability and maturity in natural ecosystems': Session P3 of *Unifying concepts in ecology. Proceedings of First International Congress of Ecology*, The Hague, 8–14 September, 1974, pp. 64–79.
Centre for Agricultural Publishing and Documentation, Wageningen, 1974.

G.W.BRYAN

Some aspects of heavy metal tolerance in aquatic organisms

In most waters the concentrations of heavy metals are very low (Riley & Chester, 1971), although higher natural concentrations occur in rivers and estuaries which are associated with outcropping metalliferous lodes. As a result, the concentrations of heavy metals in natural waters can easily be increased to levels which aquatic organisms have not previously encountered. The mechanisms possessed by organisms for handling natural fluctuations in the availability of heavy metals assume particular importance under contaminated conditions. The degree of protection afforded varies from species to species, so that under contaminated conditions the ecological balance may be tilted as the more tolerant organisms are favoured. From the public health point of view, these protective mechanisms determine the degree of contamination of edible fish and shellfish, even if the organisms themselves are unaffected.

This paper is not primarily concerned with the manifold toxic effects of heavy metals but deals with some of the processes which determine the tolerance of aquatic organisms to heavy metals.

Absorption of heavy metals

Absorption from solution

Movements of heavy metals which are attributable to active transport systems have largely been observed in unicellular organisms. In yeast, for example, arsenate is absorbed by and can ultimately inhibit the phosphate transport system (Chan & Rothstein, 1965) and zinc, cobalt and nickel are absorbed by a system which also transports magnesium and manganese (Fuhrmann & Rothstein, 1968). Absorption of nickel also appears to be an active process in the ciliate *Paramecium caudatum* (Andrivon, 1970, 1974) and in the embryo of the sea urchin *Lytechinus pictus* (Timourian & Watchmaker, 1972). It is not difficult to imagine that in higher organisms movements of heavy metals might be mediated by carrier systems used primarily for calcium or magnesium, but absorption of zinc by the liver of the puffer fish *Tetraodon hispidus* was shown to be a passive process (Saltman & Boroughs, 1960).

Pinocytosis by animal cells is quite common and has been implicated in the absorption of colloidal gold by the mantle cells of bivalve molluscs (Bevelander & Nakahara, 1966) and in the uptake of vanadium bound to mucopolysaccharides by the pharyngeal cells of ascidians (Kalk, 1963).

For the most part, the uptake of heavy metals by aquatic plants seems to be a passive process, although one which can be influenced indirectly by metabolism. Davies (1973) showed that the kinetics of zinc uptake by the diatom *Phaeodactylum tricornutum* could be explained by the rapid adsorption of zinc onto the cell membrane, followed by diffusion controlling the rate of uptake and binding to proteins within the cell. Binding to protein may control the concentration in the cell, because, during the growth cycle, the concentration of zinc reaches a maximum and then decreases as the amount of protein in the cell declines. A similar pattern has been observed for nickel in the same species (Skaar, Rystad & Jensen, 1974). It was also shown that in phosphate-starved cells the capacity for nickel absorption was low and was enhanced by pre-treatment with phosphate, presumably due to the synthesis of new binding sites.

In seaweeds, adsorption or ion-exchange processes are involved in the uptake of zinc, since the amount absorbed at different external concentrations was described by the Freundlich adsorption isotherm in the green alga *Ulva lactuca* and in the red alga *Porphyra umbilicalis* (Gutknecht, 1961, 1963, 1965). It was also shown that uptake was promoted by high pH, temperature and, in most species, light, although in the brown weed *Fucus vesiculosus* light decreased absorption. It is thought that relationships between uptake and light can be explained in terms of an indirect effect of photosynthesis on the internal pH of the plant and on the synthesis of more binding sites.

Absorption from solution by most animals seems to involve passive diffusion of the metal, probably as a soluble complex, down gradients created by adsorption at the surface (cuticle, mucus layer etc.) and binding by constituents of the surface cells, body fluids and internal organs. This pattern appears to hold for zinc in fish eggs (Wedemeyer, 1968), fish (Pentreath, 1973*a*), euphausiids (Fowler, Small & Dean, 1969), decapod crustaceans (Bryan, 1971) and polychaetes (Bryan & Hummerstone, 1973*a*). Relationships between rate of absorption and external concentration for several metals in the polychaete *Nereis diversicolor* are shown in Fig. 1. The rate of absorption of manganese is directly proportional to the external concentration, but for other metals the relationship is less exact. In the case of zinc, this is because uptake is more closely related to its adsorption on to the surface of the body during the uptake process. Direct proportionality between uptake rate and

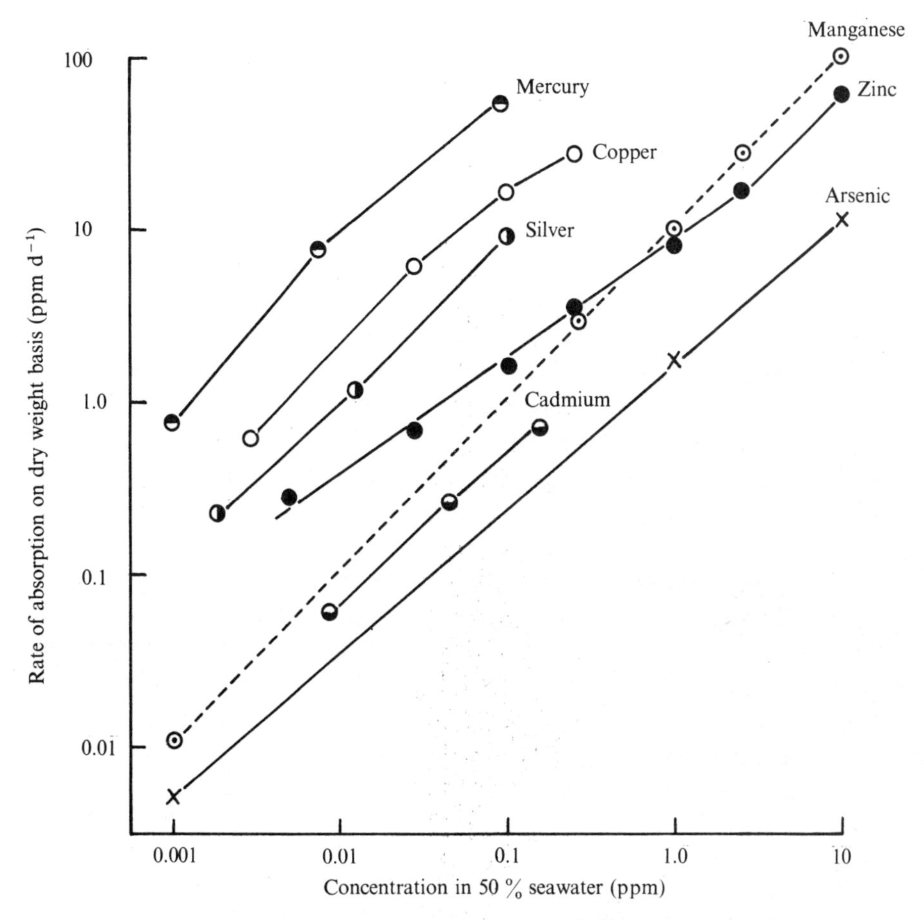

Fig. 1. Rates of absorption of metals from different concentrations in 50 % seawater by *Nereis diversicolor* from Avon estuary (Devon). Each point is a mean value for five worms and measurements were made using radioisotopes at 13 °C. Mercury was added as chloride, silver as nitrate, arsenic as arsenate and others as sulphates.

concentration has also been observed for manganese in the polychaete *Hermione hystrix* (Chipman, Schommers & Boyer, 1968), for cadmium in the shrimp *Lysmata seticaudata* and the mussel *Mytilus edulis* (Fowler & Benayoun, 1974), for chromate in the bivalve *Tapes decussatus* (Chipman, 1966) and for chromate and lead in the oyster *Crassostrea virginica* (Shuster & Pringle, 1969). There is no evidence that any animals can prevent the entry of metals by changing the permeability rapidly, although organisms such as bivalve molluscs can temporarily prevent absorption by closing the shell.

The permeability of various species is of considerable importance in determining their tolerance to metals and in *Nereis* (Fig. 1) there is a fairly close relationship between the rates of absorption of metals and their acute toxicity, metals such as copper and silver being much more toxic than zinc (see for example Tables 5, 6 and 7). Corner & Rigler (1958) showed that the much greater resistance to mercury of larvae of the brine shrimp *Artemia salina* compared with larvae of the barnacle *Elminius modestus* rested on the impermeability of the body surface in *Artemia*. It was also shown that the greater toxicity of *N*-amyl mercuric chloride to both species compared with that of mercuric chloride was due to its more rapid rate of penetration rather than to its greater toxicity to the tissues. In *Artemia* it was concluded by Corner & Sparrow (1957) that the toxicities of alkyl-mercury compounds were related to their lipid solubilities, the more lipid-soluble being able to penetrate the epidermal cell membranes more readily. This is a good example of the influence of the chemical form of a metal on its rate of absorption and toxicity. Other factors which can change the rate of absorption include changes in salinity or water hardness, the presence of other metals or of complexing agents, changes in temperature and pH, size differences and starvation. However, the tolerance of different species to heavy metals is not only determined by their rates of absorption. Changes in physiological stress may have dramatic effects on the toxicity of metals (see for example, Vernberg, Decoursey & O'Hara, 1974) and the effect of salinity on the toxicity of copper to *Nereis* (Fig. 4) is an example. Absorption of copper tends to increase with decreasing salinity and the maximum tolerance in 50 % seawater probably occurs because this estuarine species is under least stress in diluted seawater. In addition, we shall see later how tolerance also relates to the ability of organisms to excrete or detoxicate metals.

Absorption from food or particles

The evidence given in Table 1 shows that in the majority of cases food and particulates are a much more important source of metals than the water. However, most of the evidence concerns fairly large animals and was obtained with the aid of radioisotopes under normal rather than metal-contaminated conditions. Therefore it may be dangerous to generalise about the importance of different sources. Following experiments on the exchange of zinc-65 between the marine gastropod *Littorina obtusata*, its food *Fucus vesiculosus* and the surrounding water, Young (1974) concluded that 33 % of the total zinc in the mollusc was virtually non-exchangeable. Similarly, Renfro, Fowler, Heyraud & LaRosa (1974*a*) concluded from experiments with fish and crustaceans that in adult individuals refractory pools of zinc exist which exchange slowly, if at all, with zinc absorbed from water or food. This was particularly obvious in a fish, *Gobius* sp., where following the exchange of zinc-65 with food and water for 96 days a concentration factor of only 25 was achieved, compared

Table 1. *Food versus water as a source of metals*

Species	Metals	Result	Reference
Fish			
Paralichthys sp. (flounder)	Zn	Food more important than water	Hoss (1964)
Pleuronectes platessa (plaice)	Zn, Mn, Co, Fe	Water unimportant, therefore food	Pentreath (1973*a*,*b*)
Gobius sp. (goby)	Zn	Food more important than water	Renfro *et al.* (1974*a*)
Ascidians			
Ciona intestinalis (sea squirt)	V	Water most important	Kustin *et al.* (1975)
Crustaceans			
Austropotamobius pallipes (freshwater crayfish)	Zn	Food very important	Bryan (1967)
Homarus vulgaris (lobster)	Mn	Water unimportant, therefore food	Bryan & Ward (1965)
Carcinus maenas (crab)	Zn	Both sources important	Renfro *et al.* (1974*a*)
Lysmata seticaudata (shrimp)	Zn	Water more important	Renfro *et al.* (1974*a*)
Marine and intertidal isopods	Cu	Water more important	Wieser (1967)
Balanus balanoides (barnacle)	Zn, Fe	Food for Fe and mainly for Zn	Young (1974)
Molluscs			
Nucella lapillus (dogwhelk)	Zn, Fe	Nearly all from food (barnacles)	Young (1974)
Mytilus edulis (mussel)	Zn, Mn, Co, Fe	Water fairly unimportant	Pentreath (1973*c*)
Oysters	Zn, Co, Fe	Mainly food and particles	Preston & Jefferies (1969)
Octopus vulgaris (octopus)	Cu	Mainly from food (crabs)	Ghiretti & Violante (1964)

with 900 for stable zinc. Tupper, Watts & Wormall (1951) showed that zinc in the metalloenzyme carbonic anhydrase from ox erythrocytes was virtually non-exchangeable with zinc-65 *in vitro*. Thus the non-exchangeability of zinc and perhaps other metals is an indication of the presence of very stable metallic compounds in aquatic organisms.

The availability of metals from food depends on their chemical form and very stable compounds may not be broken down by digestion. Studies on the absorption of ionic zinc from the stomach in decapod crustaceans showed that it was readily absorbed and there was no evidence that uptake could be reduced significantly if high concentrations were present (Bryan, 1964, 1966, 1967, 1968). On the other hand, Pentreath (1973*a*) showed that the availability to plaice, *Pleuronectes platessa*, of zinc-65 and manganese-54 in pellets of starch or gelatine or absorbed by *Nereis* varied from 14 to 72 %. Preliminary observations suggest that copper from naturally contaminated *Nereis* is not readily absorbed by fish, although it is difficult to separate non-absorption from the ability to regulate copper and perhaps excrete it via the gut.

Excretion and regulation of heavy metals

Although rates of uptake may be related to the external concentration, there is no certainty that concentrations in the organism will reflect those of the environment. Some species are able to excrete a higher proportion of the metal intake under contaminated conditions and thereby regulate the concentration in the body at a fairly normal level.

There is not much evidence of excretory or regulatory ability in aquatic plants and perhaps losses occur mainly by diffusion. However, Mandelli (1969), who studied copper uptake in marine phytoplankton, suggested that copper may be excreted along with organic material as a detoxication mechanism. Skaar *et al.* (1974) showed that concentrations of nickel in the diatom *Phaeodactylum tricornutum* were proportional to those of seawater up to 0.75 ppm, although the concentration factor was not constant and changed with the age of the cells. Over a similar range of concentrations, direct proportionality was also observed for copper and zinc in the freshwater alga *Chlorella pyrenoidosa* (Knauss & Porter, 1954). No evidence of regulation has been found in large algae such as brown seaweeds, and in *Laminaria digitata* the relationship between water and weed for zinc is described by the Freundlich adsorption isotherm (Bryan, 1969). The concentration factor falls gradually as the level in the water increases. Under field conditions, the factor for zinc in *Fucus vesiculosus* was 6.4×10^4 (dry basis) at 0.004 ppm in the water and 1.1×10^4 at 0.113 ppm (Bryan & Hummerstone, 1973*c*). In

Fucus concentrations of metals increase markedly with distance from the growing tips. This may be due to the synthesis of new binding sites, the slow accumulation of metals and the contamination of the older surfaces with fine particles.

Apart from the possibilities of losses by diffusion, various systems for losing metals have been recognised in molluscs. These include excretion as granules from the kidneys of scallops (Bryan, 1973) and excretion in spheres pinched off from digestive cells in *Cardium edule* (Owen, 1955). In oysters, Galtsoff (1964) reported that particles of iron were ejected from the mantle edge and the importance of leucocytes in both storing and ejecting foreign materials has been the subject of several studies in oysters (Yonge, 1926; Takatsuki, 1934). Evidence that iron may be excreted via the byssus gland in *Mytilus edulis* has been given by Pentreath (1973*c*) and, in the same species, studies on the uptake of lead showed that the kidneys contained 50–70 % of the total lead and were the tissues which gained and lost it most readily (Schulz-Baldes, 1974). However, regulation of lead is poor since it was shown that the rates of uptake and the concentrations attained in 40 days were proportional to the external concentration from 0.005 to 5 ppm. It was also shown that the loss of lead in clean seawater was proportional to the internal concentration, so that the efficiency of excretion remains constant. As a result, the concentration of lead in the body rises in contaminated seawater until the rate of excretion can equal the rate of intake. Schulz-Baldes calculated that this equilibrium would take at least 230 days to achieve at a concentration factor on a dry weight basis of 3.5×10^4, assuming that equal amounts of lead were absorbed from water and food. The concentration factor for a metal often depends on the organism's size and in *Mytilus* it was shown that concentrations of lead, copper, zinc and iron fell with increasing weight whereas levels of nickel and cadmium remained constant (Boyden, 1974). These two patterns were also observed in five other species, but in the limpet *Patella vulgata* from the contaminated Bristol Channel the level of cadmium increased in proportion to the dry weight of the animal. In less contaminated areas the concentration rose much less steeply with weight. An apparently similar situation is shown in Fig. 2 for the burrowing estuarine bivalve *Scrobicularia plana*. There seems to be an upper limit to the amount of metal which can be excreted by the animal and if this is exceeded the concentration increases with size or age. So far as can be judged from the limited amount of experimental evidence and field observations (Table 2) regulation of metals is generally poor in bivalve and perhaps gastropod molluscs, but in polychaetes such as *Nereis* and *Nephthys* (Fig. 3; Table 2) zinc appears to be regulated whereas silver, lead and copper are not. In addition, some regulation of manganese and perhaps iron has also been observed in *Nereis* (Bryan & Hummerstone, 1971, 1973*b*).

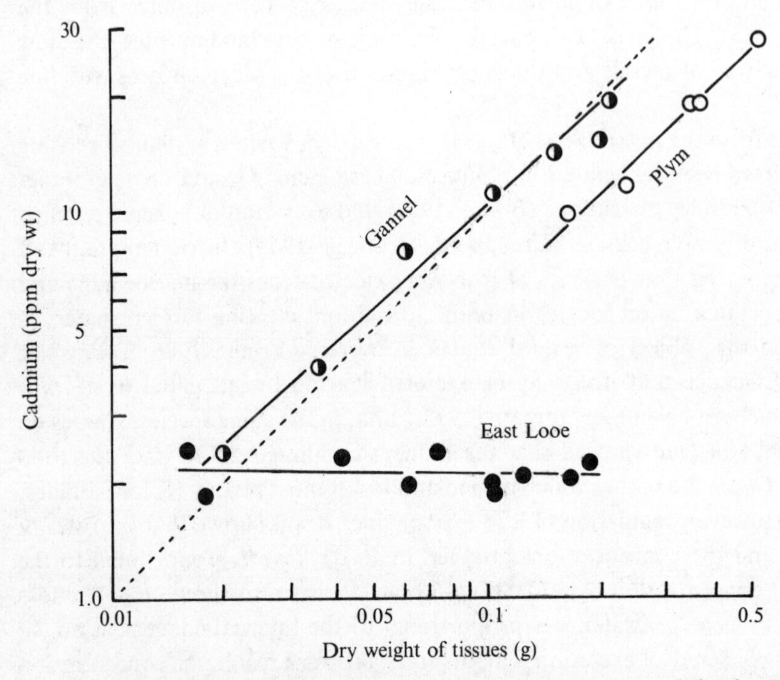

Fig. 2. Relationship between cadmium concentration and the dry weight of the soft parts in the burrowing bivalve *Scrobicularia plana* from three estuaries in south-west England. Concentration of cadmium in surface sediment was: East Looe, 0.11 ppm; Gannel, 0.52 ppm; Plym, 3.1 ppm. Each point represents a pooled sample from about five animals. Broken line shows direct proportionality.

Unlike zinc, cadmium is not regulated in *Nereis*, so that apart from copper the essential metals are regulated whereas the non-essential are not.

In decapod crustaceans, some regulation of essential metals such as zinc, copper and manganese has been observed (Bryan & Ward, 1965; Bryan, 1968) but cadmium appears not to be regulated (Table 2). Increasing the level of zinc in seawater from 0.004 to 0.2 ppm had little effect on concentrations in the lobster *Homarus vulgaris*. In this species, the urine/blood ratio of zinc was found to vary from as low as 0.05 in starved animals from low-zinc seawater to more than 4 following injections of zinc into the blood. Changes in the efficiency of zinc excretion have also been observed in the crab *Carcinus maenas*, where losses appear to take place through the gills, and in the fresh water crayfish where losses occur via the gut (Bryan, 1966, 1967, 1968, 1971).

It can be inferred from various studies with freshwater fish that zinc (Mount, 1964; Eisler, 1967), copper (Brungs, Leonard & McKim, 1973), molybdenum (Ward, 1973) and chromium (Fromm & Stokes, 1962) can be

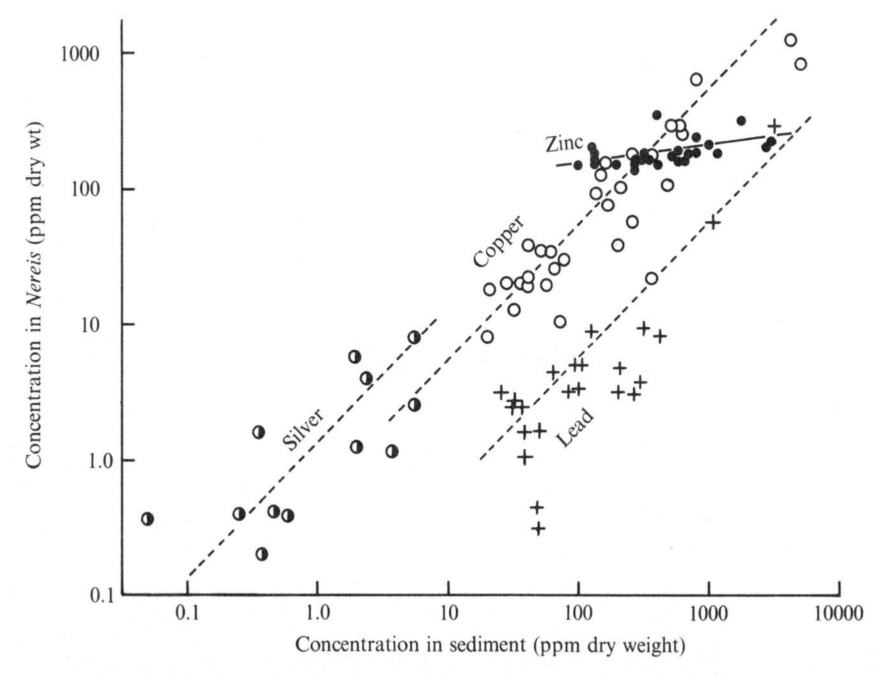

Fig. 3. Relationship between concentrations of metals in *Nereis diversicolor* and those in sediments from more than twenty estuaries in Devon and Cornwall. Each point is a mean value from an individual estuary or part of an estuary. Broken lines show direct proportionality.

regulated over a range of external concentrations. How regulation is achieved is not clear, although Nakatani (1966) showed that in the rainbow trout, *Salmo gairdneri*, under normal conditions zinc-65 was lost via the gills rather than in the urine. The above are essential metals but regulation of non-essential metals such as cadmium and sometimes mercury may occur (Table 2). Johnels *et al.* (1967) showed that low concentrations of mercury which were independent of the weight of the fish were found in pike *Esox lucius* from uncontaminated and slightly contaminated Swedish lakes. In contaminated lakes the concentration increased linearly with size, suggesting that there was an upper limit to the rate of excretion (cf. Fig. 2). In the sea, Blanton, Blanton & Robinson (1972) reported that mullet from a contaminated area retained a low level of mercury which was independent of size, suggesting that mercury may be regulated. However, even in the absence of contamination, the ability of many species of marine fish to excrete mercury is limited and concentrations in the tissues increase with size or age (Rivers, Pearson & Schultz, 1972; Cross, Hardy, Jones & Barber, 1973). This may be one of the main reasons why the highest concentrations of mercury have been found in large specimens

Table 2. *Metal concentrations (ppm dry wt) in species from chronically contaminated and normal areas*

		Copper		Zinc		Reference
Restronguet Creek, south-west England		Contaminated	Normal	Contaminated	Normal	
Platichthys flesus (flounder)	Liver	118	60	203	190	
	Viscera	28	12	123	114	
	Remainder	4.3	3.2	140	130	
Carcinus maenas (crab)	Whole	191	77	149	77	
	Hepatopancreas	2540	163	348	217	
	Gills	341	130	576	188	
Corophium volutator (amphipod)		499	96	254	130	
Nereis diversicolor (ragworm)		1140	22	194	163	
Nephthys hombergi (polychaete)		2120	18	483	305	
Scrobicularia plana (bivalve)		111	13	3240	323	
Ostrea edulis (oyster)		3870	488	14900	4130	
Fucus vesiculosus (seaweed)		2780	5	3240	90	
Derwent Estuary, Tasmania		Cadmium				
Ammotretis sp. (flounder muscle)*		< 0.25	—	66	—	Eustace (1974)
Crassostrea gigas (oyster)		173	0	57600	1390	Ratkowsky *et al.* (1974)
Bristol Channel, England						
Platichthys flesus (flounder)		~ 5	—	~ 130	—	Hardisty *et al.* (1974*b*)
Carcinus maenas (crab)*		~ 100	~ 1	~ 400	77	Peden *et al.* (1973)
Crangon vulgaris (shrimp)		125	4.9	126	101	Hardisty *et al.* (1974*a*)
Nucella lapillus (dogwhelk)		725	31	4200	175	Nickless *et al.* (1972)
Mytilus edulis (mussel)		60	4	250	62	Nickless *et al.* (1972)
Fucus vesiculosus (seaweed)		220	2	800	32	Butterworth *et al.* (1972)
Minamata Bay, Japan		Mercury				
Eight species of fish*		115	0.15	—	—	Fujiki (1973)
Venus japonica (bivalve)		178	0.3	—	—	Fujiki (1973)

* Wet weight concentration × 5.

of species such as tuna, swordfish and some sharks, which are presumably long-lived and can grow to a very large size (Cross *et al.*, 1973).

From the work discussed above, it seems fairly clear that plants and bivalve molluscs are poor regulators of heavy metals whereas essential metals such as zinc and copper are regulated in decapod crustaceans and fish. Non-essential metals such as mercury and cadmium seem to be less well regulated. The magnitude of differences between concentrations of metals in species from contaminated and more normal areas (Table 2) tends to confirm these statements.

Transformation and storage of metals

Even under normal conditions the amount of zinc required to satisfy the needs of enzyme systems is much lower than the concentrations observed in marine tissues (Pequegnat, Fowler & Small, 1969; Coombs, 1972). Under contaminated conditions the requirement for all essential metals (iron, copper, zinc, cobalt, manganese, chromium, molybdenum, vanadium, selenium, nickel and tin), must be far exceeded. Although at high concentrations both essential and non-essential metals are enzyme inhibitors, astonishingly high levels have been found in some species where regulation is poor (Table 2). How protection is achieved is not fully understood, but some of the possible mechanisms will be discussed.

Temporary protection at least is probably provided by the general binding capacity of compounds such as proteins, polysaccharides and amino acids. For example, it was found that following the injection of zinc about thirty times the normal concentration could be bound by blood proteins in the crayfish *Austropotamobius pallipes*. However, this was temporary and most of the excess was absorbed by the hepatopancreas within 2 days (Bryan, 1967). In *Procambarus clarkii*, another crayfish, copper and iron granules were found in the cells of the hepatopancreas (Ogura, 1959) and conceivably zinc is stored in the same way. In the polychaete *Nereis diversicolor*, the high concentration of copper in animals from contaminated areas (cf. Fig. 3) is stored as fine granules in the epidermal cells. These resemble the phagolysosomes pictured by Gedigk (1969) in copper-contaminated rabbit liver cells. Other examples of storage as granules are given in Table 3.

In molluscs, wandering leucocytes seem to have an important role in the translocation and detoxication of metals. This is particularly noticeable in copper-contaminated oysters where accumulation of copper by leucocytes contributes to the green colour of such animals. Studies on the binding of copper and zinc (cf. Table 2), have been carried out by Coombs (1974). He showed that in normal oysters, *Ostrea edulis*, about 60 % of the zinc was

Table 3. *Different sites for metal storage*

Species	Metals	Tissue	Comment	Reference
Mammals				
Zalophus californianus (sea lion)	Pb	Bone	Probably similar to man	Braham (1973)
Phoca vitulina (seal)	Cd	Kidney	Probably Cd-metallothionein	Heppleston & French (1973)
Halichoerus grypus (seal)	Cd	Liver	Cd-metallothionein	Olafson & Thompson (1974)
Halichoerus grypus (seal)	Hg	Liver, fur, claws	Low proportion methylmercury-demethylation?	Freeman & Horne (1973)
Delphinus delphis (dolphin)	Hg, Se	Liver	Two metals linearly correlated; not methylmercury	Koeman *et al.* (1973)
Fish				
Makaira ampla (Pacific blue marlin)	Hg	Liver, muscle	Low proportion methylmercury-demethylation?	Rivers *et al.* (1972)
Sebastodes caurinus (rock fish)	Cd	Liver	Induced Cd-metallothionein	Olafson & Thompson (1974)
Crustaceans				
Procambarus clarkii (fresh-water crayfish)	Cu, Fe	Hepatopancreas	Large granules in Fe and Cu cells	Ogura (1959)

Homarus vulgaris (lobster)	Mn	Exoskeleton		Bryan & Ward (1965)
Crangon vulgaris (shrimp)	Cu	Hepatopancreas	Excess Cu stored as granules	Djangmah (1970)
Lysmata seticaudata (shrimp)	Cd	Exoskeleton	50 % total body Cd lost at moult	Fowler & Benayoun (1974)
Molluscs				
Biomphalaria glabrata (freshwater pulmonate)	Cu	Leucocytes	Phagocytosis of excess Cu	Cheng & Sullivan (1974)
Oncomelania formosana (freshwater prosobranch)	Cu	Connective tissue	Crystals of carbonate deposited by wandering cells	Winkler & Chi (1967)
Ostrea edulis (oyster)	Cu, Zn	Leucocytes	Phago- or pinocytosis of excess metals	e.g. Boyce & Herdman (1898)
Pecten maximus (scallop)	Zn, Mn, Pb, Ag, Cd, Cu	Kidney Digestive gland	Large (5 μm) granules inside cells May occur in granules	Bryan (1973) Bryan (1973)
Octopus vulgaris (octopus)	Hg	Digestive gland	Especially in contaminated conditions	Renzoni *et al.* (1973)
Polychaete				
Nereis diversicolor (rag-worm)	Cu, Pb	Epidermis	Fine granules in high-metal worms	Bryan (1974)

firmly bound to cell debris and was readily exchangeable with zinc ions, about 18 % was bound less firmly to cell debris and was readily exchangeable and about 40 % was weakly bound to small molecular weight compounds such as taurine, lysine, ATP and possibly homarine. Compounds such as metallothionein which specifically complex metals as a method of detoxication were not observed. The presence in terrestrial mammals of storage proteins such as metallothionein for cadmium, zinc, mercury and copper (Pulido, Kagi & Vallee, 1966) and ferritin for iron has led to the search for similar proteins in aquatic organisms. Ferritin has been observed in the digestive gland of molluscs such as *Octopus vulgaris* (Nardi, Muzu & Puca, 1971) and what appears to be cadmium metallothionein has been found in the blue-green alga *Anacystis nidulans* (Maclean, Lucis, Shakh & Jansz, 1972) and in the liver of seals (Olafson & Thompson, 1974). Olafson & Thompson also reported the induction of cadmium metallothionein in the liver of the rockfish *Sebastodes caurinus* following treatment with inorganic cadmium. A similar observation was made in rabbits by Nordeberg, Piscator & Vesterberg (1972) who also observed that cadmium treatment increased the level of zinc in the liver. Therefore the possibility exists that the synthesis of compounds for detoxicating one metal may increase the affinity of the organism for another. In fact a strong correlation between concentrations of selenium and mercury has been observed in tissues from tuna (methylmercury), marine mammals and man (inorganic mercury) by Ganther & Sunde (1974), Koeman *et al.* (1973) and Kosta, Byrne & Zelenko (1975). It is suggested that the co-accumulation of selenium offers some sort of protection against mercury, but the mechanism is not known.

A high proportion of the total concentration of mercury in many fish occurs as methylmercury, the much more toxic organic form of the metal. Methylmercury can be formed by micro-organisms from divalent mercury in sediments (Jernelöv, 1972; Andren & Harriss, 1973) and this may be the ultimate source in fish, at least in freshwater. However, there is no certainty about this in marine fish and it was reported that liver homogenates from tuna could methylate inorganic mercury (Imura, Pan & Ukita, 1972). Whatever the source, the presence in some fish-eating fish and mammals of a low proportion of methylmercury suggests that demethylation to an inorganic form is occurring in the tissues (Table 3). Breakdown of methylmercury has also been observed in tissues of the killifish *Fundulus heteroclitus* (Renfro *et al.*, 1974*b*) and is known to occur in rats (Syversen, 1974). In contrast, most of the arsenic in marine plants and animals occurs in organic forms which are less toxic than inorganic arsenic. It is thought that most groups of organisms can synthesise organic from inorganic arsenic (Lunde, 1973*a,b*; Wood, 1973) and it is also thought that selenium may behave similarly (Lunde, 1970; Wood, 1973).

A considerable amount of work has been carried out on the loss of heavy metal radionuclides in the absence of metal contamination. Partly because of the incomplete exchange referred to earlier (p. 10), it is almost impossible to extrapolate such results to losses from metal-contaminated organisms. Because of the formation of stable or insoluble storage products in some cases, the tendency for metals to be lost in the absence of contamination may be very small. For example, brown seaweeds such as *Laminaria digitata* show little tendency to lose zinc (Bryan, 1969), high levels of copper and lead in contaminated *Nereis diversicolor* are not readily lost in low-metal sediments and mercury was not readily lost from the crab *Pachygrapsus marmoratus* (Renzoni *et al.*, 1973). Peden *et al.* (1973) showed that contaminated dog-whelks, *Nucella lapillus*, and crabs, *Carcinus maenas*, lost very little cadmium in clean seawater. However, these animals were starved and it could be argued that losses would have occurred had they been fed. Because the gut was the main route of excretion of zinc in the freshwater crayfish *Austropotamobius pallipes*, metal contamination was only removed when the animal was fed (Bryan, 1967).

The development of tolerance to metals

In some species increased tolerance to the toxic effects of some metals can be acquired by previous exposure to sublethal concentrations. Several examples of this have been given by Sprague (1970) for fish. Lloyd (1960) found that rainbow trout *Salmo gairdneri* were more resistant to lethal concentrations of zinc following exposure to sublethal concentrations, and in the same species Sinley, Goettl & Davies (1974) showed that more tolerant fish were produced from zinc-treated eggs. When adult brine shrimp *Artemia salina* were exposed to 0.1 ppm of copper for 3 weeks, the median lethal time in 1 ppm was approximately double that of untreated animals (Saliba & Ahsanullah, 1973). However, the same authors were unable to adapt the polychaete *Ophryotrocha labronica* to copper.

Metal-resistant strains of organisms have been collected from contaminated areas. The greater resistance to copper in strains of the brown fouling alga *Ectocarpus siliculosus* from high-copper situations was shown by Russell & Morris (1972) to be genetically determined. Some other examples in aquatic plants have been given by Whitton (1970) and Stokes, Hutchinson & Krauter (1973) isolated a nickel- and copper-resistant strain of the green alga *Scenedesmus acutiformis* from a contaminated Canadian lake.

Heavy contamination with zinc and copper occurs in Restronguet Creek, a branch of the Fal estuary in Cornwall, and species containing high concen-

Table 4. *Toxicity of copper (citrate) to tolerant and non-tolerant animals*

	96-h LC_{50} (ppm)		168-h LC_{50} (ppm)	
	Nereis diversicolor	*Nephthys hombergi**	*Scrobicularia plana*	*Corophium volutator*
Tolerant	2.3	0.7	4.6	50
Non-tolerant	0.54	0.25	2.3	32

* In 100 % seawater, others in 50 %.

trations of one or both metals have been found (Table 2). The area has been contaminated with mining wastes for more than 200 years and the surface sediments contain about 3000 ppm of copper and comparable levels of zinc and arsenic. Bryan & Hummerstone (1971) showed that *Nereis diversicolor* from this area is more resistant than normal to the acute toxicity of copper and a comparison between this and other species is shown in Table 4. *Nephthys* and *Scrobicularia* seem less well-adapted than *Nereis* but were collected in a slightly less contaminated area having a higher salinity. *Corophium* seems to be inherently resistant to copper and was found in situations where *Nereis* was unable to survive. Seaweeds may also be adapted, since fronds of *Fucus vesiculosus* from this area appeared unaffected by 1 ppm of copper (citrate) in 50 % seawater, which caused the disintegration of weed from a low-copper estuary.

Experiments with Nereis diversicolor

Having recognised from a survey of estuaries in Devon and Cornwall which populations contained high or low concentrations of metals (Fig. 3), toxicity experiments were carried out as a means of comparing their tolerance to metals.

Toxicity experiments. Great care was taken to compare animals of similar sizes and a low-metal population from the Avon estuary (Devon) was always used simultaneously as a reference. Having been acclimatised to 50 % seawater for 4–5 days, groups of ten worms weighing about 0.3 g each were exposed to different concentrations of metal in 50 % seawater at 13 °C. The 500 ml of solution was aerated and changed daily. When an animal no longer reacted to mechanical stimulation it was removed from the experiment.

Table 5 shows the median lethal times in copper solutions for five populations containing different concentrations of copper. The following points are evident.

Table 5. *Toxicity of copper (citrate) in 50% seawater to* Nereis

Estuary	Sediment: typical Cu (ppm)	Worms: initial Cu (ppm dry)	No. of expts	Median lethal times* 0.5 ppm	1.0 ppm	2.5 ppm	Mean %
Avon	18	20	4	100 (155 h)	100 (59 h)	100 (27 h)	100
Gannel	296	116	2	128	100	135	121
Tamar	509	397	2	97	93	111	100
Hayle	712	729	3	490+	230	260	327
Restronguet Creek	3500	922	3	440+	299	259	333

* As percentage of Avon controls at three concentrations.

(1) The high-copper populations from Hayle and Restronguet Creek are both tolerant to copper, although they are geographically separated by being on the north and south sides of the Cornish peninsula.

(2) Although the Tamar population contains an appreciable amount of copper it is not particularly resistant, suggesting that tolerance is not simply related to the amount of copper in the worm.

(3) The Gannel population contained more than 600 ppm of lead, which is more than 100 times the levels in other populations (cf. Table 10). Thus a possible tolerance to lead does not confer on the worms an appreciable resistance to copper.

Average salinities of the sediments from which these worms were collected varied with time of year, but 50 % seawater, as used in the experiments, provided a reasonable compromise. Fig. 4 shows that the greater tolerance of the Restronguet Creek population is maintained over a wide range of salinities.

Zinc is much less toxic than copper and seems to be regulated in the body by animals from a wide range of sediments (Fig. 3; Table 6). Although the Hayle and Restronguet Creek populations are resistant to copper, only the latter from the most heavily contaminated sediment is resistant to zinc (Table 6). Worms from the Gannel estuary are only slightly more resistant to zinc despite the high concentration in the sediment and the fact that the worms contain high concentrations of lead. Therefore, there is no evidence that tolerance to copper or possibly lead predisposes worms to be tolerant to zinc.

Lead concentrations exceeding 8000 ppm in the sediments and 1000 ppm in the worms have been observed in parts of the Gannel estuary. However, toxicity experiments comparable with those for copper and zinc have not been possible because the solubility of lead is less than 5 ppm in 50 % seawater and this does not seem to be toxic.

Silver also poses solubility problems, but in several experiments using

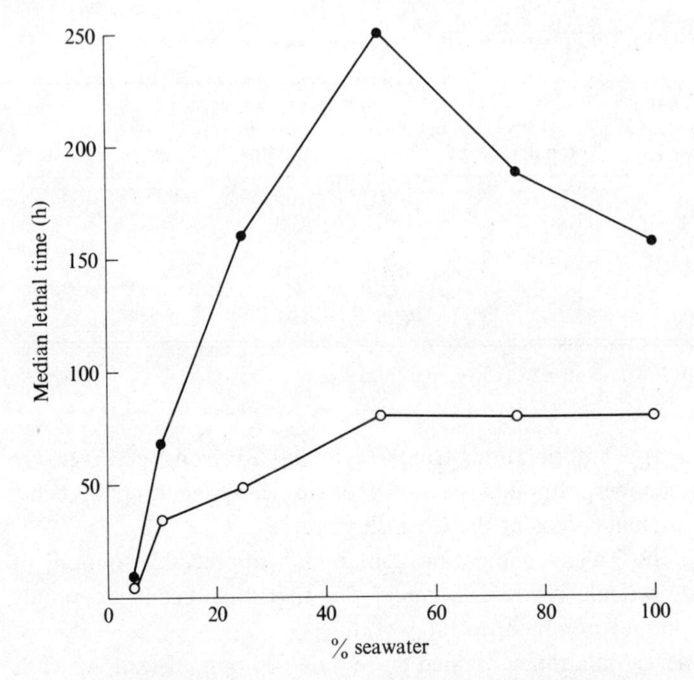

Fig. 4. Effect of salinity on the median lethal time for tolerant (●, Restronguet Creek) and non-tolerant (○, Avon) populations of *Nereis diversicolor* in 1 ppm copper (citrate) at 13 °C.

0.5 ppm it was shown conclusively that the high-copper animals from Hayle and Restronguet Creek and the high-lead animals from the Gannel estuary were more tolerant than the Avon and East Looe animals (Table 7). The East Looe population was involved because these animals contain as much silver as the three more tolerant populations. Therefore, it would seem that resistance to silver depends on the presence of high concentrations of other metals rather than on the presence of silver itself. During these experiments it was seen that, starting with the head end, the dorsal surface of some animals became grey or white in colour. This ranged from very obvious in the Restronguet Creek worms to slight in some East Looe animals and hardly at all in the Avon animals. Sections of the Restronguet Creek worms showed that the cuticle (or possibly the microvilli which penetrate from the epidermal cells to the surface) was heavily stained with silver and the epidermal cells were filled with stained granules which resemble those staining for copper in the same population (cf. Table 3).

Absorption experiments. By studying the rate of uptake of zinc-65 from 50 % seawater containing 0.01, 0.1 and 1.0 ppm of added zinc, it was shown that

Table 6. *Toxicity of zinc (sulphate) in 50 % seawater to* Nereis

Estuary	Sediment: typical Zn (ppm)	Worms: initial Zn (ppm dry)	No. of expts	Median lethal times* 25 ppm	50 ppm	Mean %
Avon	92	151	2	100 (264 h)	100 (96 h)	100
Hayle	895	195	2	72	80	76
Gannel	2010	215	2	131	107	119
Restronguet Creek	2960	204	2	173	179	176

* As percentage of Avon control at two concentrations.

Table 7. *Toxicity of silver (nitrate) in 50 % seawater to* Nereis

Estuary	Sediment: typical Ag (ppm)	Worms: initial Ag (ppm dry)	No. of expts	Median lethal time*
Avon	< 0.2	0.7	2	100 (180 h)
Hayle	1.5	5.7	2	160+ unaffected
East Looe	3.1	5.0	2	143
Restronguet Creek	4.6	5.2	2	160+ unaffected
Gannel	13.6	10.8	2	160+ slight effect

* As percentage of Avon control in 0.5 ppm.

the zinc-tolerant animals from Restronguet Creek were 30–35 % less permeable to the metal than non-tolerant Avon animals (Bryan & Hummerstone, 1973*a*). At higher non-toxic concentrations of 5 and 10 ppm, the net uptake of stable zinc was 43–50 % lower in the tolerant worms and seems to be the main reason for their resistance to acute toxicity. This lower permeability does not seem to be specific to zinc, since it was also found that the Restronguet Creek population was about 15 % less permeable to cadmium and 35 % less permeable to manganese.

Using copper-64 it was found that, unlike zinc, the metal was absorbed more rapidly by the tolerant than the non-tolerant animals (Bryan, 1974). It was thought that this might be a feature of copper-tolerant worms, reflecting their greater capacity for binding and detoxicating copper in the epidermal cells. When, however, a comparison was made between all five populations from the toxicity experiments in Table 5, it was found that copper was rapidly absorbed by the Tamar population which, although non-tolerant, contains a

Table 8. *Rates of absorption of copper ($^{64}CuSO_4$) from different concentrations in 50 % seawater over period 14–110 h*

Estuary	Worms: initial Cu (ppm dry)	Mean rates using 5 worms at each concentration (ppm d^{-1} on dry basis)*				
		0.0029 ppm	0.028 ppm	0.106 ppm	0.256 ppm	1.01 ppm
Avon	22	0.60	6.2	16.3	26.8	—
Gannel	145	1.27	—	14.8	—	—
Tamar	508	2.38	—	35.1	—	—
Hayle	1003	2.16	—	9.2	—	—
Restronguet Creek	901	2.24	15.2	21.9	37.3	105

* Divide by seven for rate on wet basis.

Table 9. *Comparison of copper toxicity and rates of copper uptake in 50 % seawater by fresh-collected and laboratory-grown* Nereis

Estuary	Type	Worms: initial Cu (ppm dry)	Median lethal time in 1 ppm $CuSO_4$ as % Avon fresh	Mean rates of uptake using 5 worms per concentration (ppm d^{-1} on dry basis)	
				0.0034 ppm	0.028 ppm
Avon	Fresh	16	100 (48 h)	1.32	9.35
Avon	Grown	11	137	0.58	8.82
Restronguet Creek	Fresh	882	306	2.55	34.3
Restronguet Creek	Grown	99	271	0.54	4.34

high concentration of copper. Thus the rapid absorption of copper from non-toxic concentrations seems to reflect the presence of material for binding the metal but does not necessarily reflect the existence of copper tolerance.

To examine this point in more detail, tiny animals from the Avon estuary and Restronguet Creek were reared for 6 months on yeast in acid-washed sand covered with 50 % seawater until adult size was reached. Then, the toxicity of copper and the rates of absorption of copper-64 by the grown animals were compared with those of fresh animals of equal size. Table 9 shows that the level of copper in the laboratory-grown Restronguet Creek animals was 9 times lower than in the fresh worms, but the tolerance to copper was reduced only slightly. Again this suggests that tolerance is not necessarily related to the

Table 10. *Rates of absorption of silver* (${}^{110}AgNO_3$) *from different concentrations in* 50 % *seawater over period* 70–200 *h*

	Initial concentrations in worms (ppm dry)			Mean rates using 5 worms at each concentration (ppm d^{-1} on dry basis)*		
Estuary	Ag	Cu	Pb	0.0019 ppm	0.012 ppm	0.10 ppm
Avon	0	19	4.1	0.22	1.14	9.08
East Looe	4.2	24	4.0	0.30	0.74	6.21
Gannel	7.1	103	1034	0.91	3.14	9.41
Hayle	4.7	631	2.7	0.69	3.32	12.9
Restronguet Creek	2.6	792	3.5	1.01	7.05	36.2

* Divide by seven for rate on wet basis.

Table 11. *Rates of absorption of mercury* (${}^{203}HgCl_2$) *from concentrations in* 50 % *seawater over period* 70–240 *h*

		Mean rates using 5 worms at each concentration (ppm d^{-1} on dry basis)*		
Estuary	Type	0.00096 ppm	0.0073 ppm	0.093 ppm
Avon	Low Cu	0.78	7.26	51.6
Hayle	High Cu	0.80	10.8	51.0
Restronguet Creek	High Cu	0.86	7.7	64.5

* Divide by seven for rate on wet basis.

presence of a high concentration of copper in the animal. Rates of absorption of copper from low external concentrations were reduced in the laboratory-grown Restronguet Creek animals to values below those for the Avon animals. This suggests that the persistent tolerance of the Restronguet Creek worms could have its basis in a low permeability to copper which is normally masked by the greater tendency for high-copper animals to absorb and bind copper in the epidermal cells (cf. Table 8). Possibly the low permeability in the laboratory-grown Restronguet Creek animals and that to other metals such as zinc in the fresh-collected animals is a function of the inner cell membranes of the epidermal cells rather than the outer membranes. This would not reduce the rate of penetration of metals into the cells from the water but would reduce the rate of entry to other more vulnerable tissues.

The rates of absorption of silver from three external concentrations by the populations used in the toxicity experiments in Table 7 are shown in Table 10. At the two lower concentrations silver is absorbed more rapidly by the more tolerant high-lead and high-copper populations. At 0.10 ppm the differences become less obvious, but the capacity for uptake remains strikingly high in the Restronguet Creek worms. These results may indicate that the excess metal-binding capacity of the epidermal cells is being titrated with silver and the same may perhaps be said of the uptake of copper in the previous section. Whether the tolerance to silver in some populations depends on this binding capacity or whether underlying this there is a lower permeability to silver as was indicated for copper in the Restronguet Creek animals, is at present uncertain. When similar experiments were carried out using mercury (Table 11), much smaller differences were found between the high-copper and low-copper populations than were found for silver. Possibly this reflects the greater chemical similarity between copper and silver.

To sum up: Tolerance in *Nereis* has been observed to zinc, copper, silver and possibly lead. The tolerances to zinc, copper and perhaps lead seem to have developed separately, but that to silver depends on the presence of tolerance to copper or perhaps lead. Whether tolerance is genetically determined is not known, although circumstantial evidence and the general adaptability of estuarine organisms would suggest that it is. Because metals such as copper are more toxic at low salinities (Fig. 4), the pressure to adapt is probably greatest under these conditions.

From the point of view of the organism and the polluters, the ability to adapt to metals seems to be a good thing. However, it should be remembered that many of these tolerant organisms contain two to three orders of magnitude higher concentrations of metals than normal and, so far as we know at present, these may be transmitted to non-adapted predators, including man.

Some of the research described in this paper was carried out with the financial assistance of the Department of the Environment.

References

Andren, A. W. & Harriss, R. C. (1973). Methylmercury in estuarine sediments. *Nature, London*, **245**, 256–7.

Andrivon, C. (1970). Preuves de l'existence d'un transport actif de l'ion nickel à travers la membrane cellulaire de *Paramecium caudatum*. *Protistologica*, **6**, 445–55.

Andrivon, C. (1974). La perméabilité à l'ion nickel chez *Paramecium*: ses rapports avec le renversement ciliaire. *Protistologica*, **10**, 175–83.

Bevelander, G. & Nakahara, H. (1966). Correlation of lysosomal activity and ingestion by mantle epithelium. *Biological Bulletin, Woods Hole*, **131**, 76–82.

Blanton, W. G., Blanton, C. J. & Robinson, M. C. (1972). *The ecological impact of mercury discharge on an enclosed secondary bay*. Forth Worth, Texas, Environmental Monitors. 231 pp.

Boyce, R. & Herdman, W. A. (1898). On a green leucocytosis in oysters associated with the presence of copper in leucocytes. *Proceedings of the Royal Society, London*, **62**, 30–8.

Boyden, C. R. (1974). Trace element content and body size in molluscs. *Nature, London*, **251**, 311–14.

Braham, H. W. (1973). Lead in the California sea lion (*Zalophus californianus*). *Environmental Pollution*, **5**, 253–8.

Brungs, W. A., Leonard, E. N. & McKim, J. M. (1973). Acute and long-term accumulation of copper by the brown bullhead, *Ictalurus nebulosus*. *Journal of the Fisheries Research Board of Canada*, **30**, 583–6.

Bryan, G. W. (1964). Zinc regulation in the lobster *Homarus vulgaris*. I. Tissue zinc and copper concentrations. *Journal of the Marine Biological Association of the United Kingdom*, **44**, 549–63.

Bryan, G. W. (1966). The metabolism of Zn and ^{65}Zn in crabs, lobsters and fresh-water crayfish. In *Radioecological concentration processes* (ed. B. Aberg & F. P. Hungate), pp. 1005–16. Pergamon Press, Oxford.

Bryan, G. W. (1967). Zinc regulation in the fresh-water crayfish (including some comparative copper analyses). *Journal of Experimental Biology*, **46**, 281–96.

Bryan, G. W. (1968). Concentrations of zinc and copper in the tissues of decapod crustaceans. *Journal of the Marine Biological Association of the United Kingdom*, **48**, 303–21.

Bryan, G. W. (1969). The absorption of zinc and other metals by the brown seaweed *Laminaria digitata*. *Journal of the Marine Biological Association of the United Kingdom*, **49**, 225–43.

Bryan, G. W. (1971). The effects of heavy metals (other than mercury) on marine and estuarine organisms. *Proceedings of the Royal Society, London*, Ser. B, **177**, 389–410.

Bryan, G. W. (1973). The occurrence and seasonal variation of trace metals in the scallops *Pecten maximus* (L.) and *Chlamys opercularis* (L.). *Journal of the Marine Biological Association of the United Kingdom*, **53**, 145–66.

Bryan, G. W. (1974). Adaptation of an estuarine polychaete to sediments containing high concentrations of heavy metals. In *Pollution and physiology of marine organisms* (ed. F. J. Vernberg & W. B. Vernberg), pp. 123–35. Academic Press, New York & London.

Bryan, G. W. & Hummerstone, L. G. (1971). Adaptation of the polychaete *Nereis diversicolor* to estuarine sediments containing high concentrations of heavy metals. I. General observations and adaptation to copper. *Journal of the Marine Biological Association of the United Kingdom*, **51**, 845–63.

Bryan, G. W. & Hummerstone, L. G. (1973*a*). Adaptation of the polychaete *Nereis diversicolor* to estuarine sediments containing high concentrations of zinc and cadmium. *Journal of the Marine Biological Association of the United Kingdom*, **53**, 839–57.

Bryan, G. W. & Hummerstone, L. G. (1973*b*). Adaptation of the polychaete *Nereis diversicolor* to manganese in estuarine sediments. *Journal of the Marine Biological Association of the United Kingdom*, **53**, 859–72.

Bryan, G. W. & Hummerstone, L. G. (1973*c*). Brown seaweed as an indicator of heavy metals in estuaries in south-west England. *Journal of the Marine Biological Association of the United Kingdom*, **53**, 705–20.

Bryan, G. W. & Ward, E. (1965). The absorption and loss of radioactive and non-radioactive manganese by the lobster, *Homarus vulgaris. Journal of the Marine Biological Association of the United Kingdom*, **45**, 65–95.

Butterworth, J., Lester, P. & Nickless, G. (1972). Distribution of heavy metals in the Severn Estuary. *Marine Pollution Bulletin*, **3**, 72–4.

Chan, J. & Rothstein, A. (1965). Arsenate uptake and release in relation to inhibition of transport and glycolysis in yeast. *Biochemical Pharmacology*, **14**, 1093–112.

Cheng, T. C. & Sullivan, J. T. (1974). Mode of entry, action and toxicity of copper molluscicides. In *Molluscicides in schistosomiasis control* (ed. T. C. Cheng), pp. 89–153. Academic Press, New York & London.

Chipman, W. A. (1966). Uptake and accumulation of chromium-51 by the clam, *Tapes decussatus*, in relation to physical and chemical form. In *Disposal of radioactive wastes into seas, oceans and surface waters*, pp. 571–82. International Atomic Energy Agency, Vienna.

Chipman, W., Schommers, E. & Boyer, M. (1968). Uptake, accumulation and retention of radioactive manganese by the marine annelid worm, *Hermione hystrix. Radioactivity in the sea, No. 25*. International Atomic Energy Agency, Vienna.

Coombs, T. L. (1972). The distribution of zinc in the oyster *Ostrea edulis* and its relation to enzymic activity and to other metals. *Marine Biology*, **12**, 170–8.

Coombs, T. L. (1974). The nature of zinc and copper complexes in the oyster *Ostrea edulis. Marine Biology*, **28**, 1–10.

Corner, E. D. S. & Rigler, F. H. (1958). The modes of action of toxic agents. III. Mercuric chloride and *N*-amylmercuric chloride on crustaceans. *Journal of the Marine Biological Association of the United Kingdom*, **37**, 85–96.

Corner, E. D. S. & Sparrow, B. W. (1957). The modes of action of toxic agents. II. Factors influencing the toxicities of mercury compounds to certain crustacea. *Journal of the Marine Biological Association of the United Kingdom*, **36**, 459–72.

Cross, F. A., Hardy, L. H., Jones N. Y. & Barber, R. (1973). Relation between total body weight and concentrations of manganese, iron, copper, zinc and mercury in white muscle of bluefish (*Pomatomus saltatrix*) and a bathyl-demersal fish *Antimora rostrata. Journal of the Fisheries Research Board of Canada*, **30**, 1287–91.

Davies, A. G. (1973). In *Radioactive contamination of the marine environment*, pp. 403–20. International Atomic Energy Agency, Vienna.

Djangmah, J. S. (1970). The effects of feeding and starvation on copper in the blood and hepatopancreas and on the blood proteins of *Crangon vulgaris*. (Fabricius). *Comparative Biochemistry and Physiology*, **32**, 709–31.

Eisler, R. (1967). Acute toxicity of zinc to the killifish, *Fundulus heteroclitus*. *Chesapeake Science*, **8**, 262–4.

Eustace, I. J. (1974). Zinc, cadmium, copper and manganese in species of finfish and shellfish caught in the Derwent Estuary, Tasmania. *Australian Journal of Marine and Freshwater Research*, **25**, 209–20.

Fowler, S. W. & Benayoun, G. (1974). Experimental studies on cadmium flux through marine biota. *Radioactivity in the sea, No. 44*. International Atomic Energy Agency, Vienna.

Fowler, S. W., Small, L. F. & Dean, J. M. (1969). Metabolism of zinc-65 in euphausiids. In *Symposium on radioecology* (ed. D. F. Nelson & F. C. Evans), pp. 399–411. CONF-670503, US Atomic Energy Commission.

Freeman, H. C. & Horne, D. A. (1973). Mercury in Canadian seals. *Bulletin of Environmental Contamination and Toxicology*, **10**, 172–80.

Fromm, P. O. & Stokes, R. M. (1962). Assimilation and metabolism of chromium by trout. *Journal of the Water Pollution Control Federation*, **34**, 1151–5.

Fuhrmann, G. & Rothstein, A. (1968). Transport of Zn^{2+}, Co^{2+} and Ni^{2+} into yeast cells. *Biochimica et Biophysica Acta*, **163**, 325–30.

Fujiki, M. (1973). The transitional condition of Minamata Bay and the neighbouring sea polluted by factory waste water containing mercury. In *Advances in water pollution research, Proceedings of the 6th International Conference*, Jerusalem, June 18–23, pp. 905–20. Pergamon Press, Oxford.

Galtsoff, P. S. (1964). The American oyster. *Fishery Bulletin of the Fish and Wildlife Service, US Department of the Interior*, **64**, 383–96.

Ganther, H. E. & Sunde, M. L. (1974). Effect of tuna fish and selenium on the toxicity of methylmercury: a progress report. *Journal of Food Science*, **39**, 1–5.

Gedigk, P. (1969). Pigmentation caused by inorganic materials. In *Pigments in pathology* (ed. M. Wolman), pp. 1–32. Academic Press, New York & London.

Ghiretti, F. & Violante, U. (1964). Richerche sul metabolismo del rame in *Octopus vulgaris*. *Bollettino di Zoologia*, **31**, 1081–92.

Gutknecht, J. (1961). Mechanism of radioactive zinc uptake by *Ulva lactuca*. *Limnology and Oceanography*, **6**, 426–31.

Gutknecht, J. (1963). ^{65}Zn uptake by benthic marine algae. *Limnology and Oceanography*, **8**, 31–8.

Gutknecht, J. (1965). Uptake and retention of cesium-137 and zinc-65 by seaweeds. *Limnology and Oceanography*, **10**, 58–66.

Hardisty, M. W., Huggins, R. J., Kartar, S. & Sainsbury, M. (1974*a*). Ecological implications of heavy metals in fish from the Severn Estuary. *Marine Pollution Bulletin*, **5**, 12–15.

Hardisty, M. W., Katar, S. & Sainsbury, M. (1974*b*). Dietary habits and heavy metal concentrations in fish from the Severn Estuary and Bristol Channel. *Marine Pollution Bulletin*, **5**, 61–3.

Heppleston, P. B. & French, M. C. (1973). Mercury and other metals in British seals. *Nature, London*, **243**, 302–4.

Hoss, D. E. (1964). Accumulation of zinc-65 by flounder of the genus *Paralichthys*. *Transactions of the American Fisheries Society*, **93**, 364–8.

Imura, N., Pan, S. K. & Ukita, T. (1972). Methylation of inorganic mercury with liver homogenate of tuna fish. *Chemosphere*, **1**, 197–201.

Jernelöv, A. (1972). Factors in the transformation of mercury to methylmercury. In *Environmental mercury contamination* (ed. R. Hartung & B. D. Dinman), pp. 167–72. Ann Arbour Science Publishers, Inc., Michigan.

Johnels, A. G., Westermark, T., Berg, W., Persson, P. I. & Sjöstrand, B. (1967). Pike (*Esox lucius* L.) and some other aquatic organisms in Sweden as indicators of mercury contamination in the environment. *Oikos*, **18**, 323–33.

Kalk, M. (1963). Absorption of vanadium by tunicates. *Nature, London*, **198**, 1010–11.

Knauss, H. J. & Porter, J. W. (1954). The absorption of inorganic ions by *Chlorella pyrenoidosa*. *Plant Physiology*, **29**, 229–34.

Koeman, J. H., Peeters, W. H. M., Koudstaal-Hol, C. H. M., Tjioe, P. S. & Degoeij, J. J. M. (1973). Mercury–selenium correlations in marine mammals. *Nature, London*, **245**, 385–6.

Kosta, L., Byrne, A. R. & Zelenko, V. (1975). Correlation between selenium and mercury in man following exposure to inorganic mercury. *Nature, London*, **254**, 238–9.

Kustin, K., Ladd, K. V. & McLeod, G. C. (1975). Site and rate of vanadium assimilation in the tunicate *Ciona intestinalis*. *Journal of General Physiology*, **65**, 315–28.

Lloyd, R. (1960). The toxicity of zinc sulphate to rainbow trout. *Annals of Applied Biology*, **48**, 84–94.

Lunde, G. (1970). Analysis of arsenic and selenium in marine raw materials. *Journal of the Science of Food and Agriculture*, **21**, 242–7.

Lunde, G. (1973*a*). The synthesis of fat and water soluble arseno organic compounds in marine and limnetic algae. *Acta Chemica Scandinavica*, **27**, 1586–94.

Lunde, G. (1973*b*). Separation and analysis of organic-bound and inorganic arsenic in marine organisms. *Journal of the Science of Food and Agriculture*, **24**, 1021–7.

MacLean, F. I., Lucis, O. J., Shakh, Z. A. & Jansz, E. R. (1972). The uptake and subcellular distribution of Cd and Zn in microorganisms. *Proceedings, Federation of American Societies for Experimental Biology*, **31**, 699.

Mandelli, E. F. (1969). The inhibitory effects of copper on marine phytoplankton. *Contributions in Marine Science, University of Texas*, **14**, 47–57.

Mount, D. I. (1964). An autopsy technique for zinc-caused fish mortality. *Transactions of the American Fisheries Society*, **93**, 174–82.

Nakatani, R. E. (1966). Biological response of rainbow trout (*Salmo gairdneri*) ingesting zinc-65. In *Disposal of radioactive wastes in seas, oceans and surface waters*, pp. 809–23. International Atomic Energy Agency, Vienna.

Nardi, G., Muzu, E. O. & Puca, M. (1971). Ferritin in the hepatopancreas of *Octopus vulgaris* Lam. *Comparative Biochemistry and Physiology*, **40B**, 199–205.

Nickless, G., Stenner, R. & Terrille, N. (1972). Distribution of cadmium, lead and zinc in the Bristol Channel. *Marine Pollution Bulletin*, **3**, 188–90.

Nordeberg, M., Piscator, M. & Vesterberg, O. (1972). Separation of two forms of rabbit metallothionein by isoelectric focusing. *Biochemical Journal*, **126**, 491–8.

Ogura, K. (1959). Midgut gland cells accumulating iron or copper in the crayfish, *Procambarus clarkii*. *Annotationes Zoologicae Japonenses*, **32**, 133–42.

Olafson, R. W. & Thompson, J. A. J. (1974). Isolation of heavy metal binding proteins from marine vertebrates. *Marine Biology*, **28**, 83–6.

Owen, G. (1955). Observations on the stomach and digestive diverticula of the Lamellibranchia. I. The Anisomyaria and Eulamellibranchia. *Quarterly Journal of Microscopical Science*, **96**, 517–37.

Peden, J. D., Crothers, J. H., Waterfall, C. E. & Beasley, J. (1973). Heavy metals in Somerset marine organisms. *Marine Pollution Bulletin*, **4**, 7–9.

Pentreath, R. J. (1973*a*). The accumulation and retention of ^{65}Zn and ^{54}Mn by the plaice, *Pleuronectes platessa* L. *Journal of Experimental Marine Biology and Ecology*, **12**, 1–18.

Pentreath, R. J. (1973*b*). The accumulation and retention of ^{59}Fe and ^{58}Co by the plaice, *Pleuronectes platessa* L. *Journal of Experimental Marine Biology and Ecology*, **12**, 315–26.

Pentreath, R. J. (1973*c*). The accumulation from water of ^{65}Zn, ^{54}Mn, ^{58}Co and ^{59}Fe by the mussel, *Mytilus edulis*. *Journal of the Marine Biological Association of the United Kingdom*, **53**, 127–43.

Pequegnat, J. E., Fowler, S. W. & Small, L. F. (1969). Estimates of the zinc requirements of marine organisms. *Journal of the Fisheries Research Board of Canada*, **26**, 145–50.

Preston, A. & Jefferies, D. F. (1969). Aquatic aspects in chronic and acute con-

tamination situations. In *Environmental contamination by radioactive materials*, pp. 183–211. International Atomic Energy Agency, Vienna.
Pulido, P., Kagi, J. H. R. & Vallee, B. L. (1966). Isolation and some properties of human metallothionein. *Biochemistry, New York*, **5**, 1768–77.
Ratkowsky, D. A., Thrower, S. J., Eustace, I. J. & Olley, J. (1974). A numerical study of the concentration of some heavy metals in Tasmanian oysters. *Journal of the Fisheries Research Board of Canada*, **31**, 1165–71.
Renfro, W. C., Fowler, S. W., Heyraud, M. & Larosa, J. (1974). *Relative importance of food and water pathways in the bio-accumulation of zinc. Technical Report IAEA-163*, pp. 11–20. International Atomic Energy Agency, Vienna.
Renfro, J. L., Schmidt-Nielsen, B., Miller, D., Benos, D. & Allen, J. (1974). Methyl mercury and inorganic mercury: uptake, distribution and effect on osmoregulatory mechanisms in fishes. In *Pollution and physiology of marine organisms* (ed. F. J. Vernberg & W. B. Vernberg), pp. 101–22. Academic Press, New York & London.
Renzoni, A., Bacci, E. & Falciai, L. (1973). Mercury concentration in the water, sediments and fauna of an area of the Tyrrhenian coast. *Revue Internationale d'Océanographie Médicale*, **31/2**, 17–45.
Riley, J. P. & Chester, R. (1971). *Introduction to marine chemistry*. Academic Press, New York & London.
Rivers, J. B., Pearson, J. E. & Schultz, C. D. (1972). Total and organic mercury in marine fish. *Bulletin of Environmental Contamination and Toxicology*, **8**, 257–66.
Russell, G. & Morris, O. P. (1972). Ship-fouling as an evolutionary process. *Proceedings of the third International Congress on Marine Corrosion and Fouling*, Washington, October 2–9, 1972, pp. 719–30.
Saliba, L. J. & Ahsanullah, M. (1973). Acclimation and tolerance of *Artemia salina* and *Ophryotrocha labronica* to copper sulphate. *Marine Biology*, **23**, 297–302.
Saltman, P. & Boroughs, H. (1960). The accumulation of zinc by fish liver slices. *Archives of Biochemistry and Biophysics*, **86**, 169–74.
Schulz-Baldes, M. (1974). Lead uptake from sea water and food, and lead loss in the common mussel *Mytilus edulis*. *Marine Biology*, **25**, 177–93.
Shuster, C. N. & Pringle, B. H. (1969). Trace metal accumulation by the American eastern oyster, *Crassostrea virginica*. *Proceedings of the National Shellfisheries Association*, **59**, 91–103.
Sinley, J. R., Goettl, J. P. & Davies, P. H. (1974). The effects of zinc on rainbow trout (*Salmo gairdneri*) in hard and soft water. *Bulletin of Environmental Contamination and Toxicology*, **12**, 193–201.
Skaar, H., Rystad, B. & Jensen, A. (1974). The uptake of ^{63}Ni by the diatom *Phaeodactylum tricornutum*. *Physiologia Plantarum*, **32**, 353–8.
Sprague, J. B. (1970). Measurement of pollutant toxicity to fish. II. Utilizing and applying bioassay results. *Water Research*, **4**, 3–32.
Stokes, P. M., Hutchinson, T. C. & Krauter, K. (1973). Heavy-metal tolerance in algae isolated from contaminated lakes near Sudbury, Ontario. *Canadian Journal of Botany*, **51**, 2155–68.
Syversen, T. L. M. (1974). Biotransformation of Hg-203 labelled methyl mercuric chloride in rat brain measured by specific determination of Hg^{2+}. *Acta Pharmacologica et Toxicologica*, **35**, 277–83.
Takatsuki, S. (1934). On the nature and functions of the amoebocytes of *Ostrea edulis*. *Quarterly Journal of Microscopical Science*, **76**, 379–431.
Timourian, H. & Watchmaker, G. (1972). Nickel uptake by sea urchin embryos

and their subsequent development. *Journal of Experimental Zoology*, **182**, 379–88.

Tupper, R., Watts, R. W. E. & Wormall, A. (1951). Some observations on the zinc in carbonic anhydrase. *Biochemical Journal*, **50**, 429–32.

Vernberg, W. B., Decoursey, P. & O'Hara, J. (1974). Multiple environmental effects on physiology and behaviour of the fiddler crab, *Uca pugilator*. In *Pollution and physiology of marine organisms* (ed. J. F. Vernberg & W. B. Vernberg), pp. 381–425. Academic Press, New York & London.

Ward, J. V. (1973). Molybdenum concentrations in tissues of the rainbow trout (*Salmo gairdneri*) and kokanee salmon (*Oncorhynchus nerka*) from waters differing widely in molybdenum content. *Journal of the Fisheries Research Board of Canada*, **30**, 841–2.

Wedemeyer, G. (1968). Uptake and distribution of Zn^{65} in the coho salmon egg (*Oncorhynchus kisutch*). *Comparative Biochemistry and Physiology*, **26**, 271–9.

Whitton, B. A. (1970). Toxicity of heavy metals to freshwater algae: a review. *Phycos*, **9**, 116–25.

Wieser, W. (1967). Conquering terra firma: The copper problem from the isopod's point of view. *Helgoländer Wissenschaftliche Meersuntersuchungen*, **15**, 282–93.

Winkler, L. R. & Chi, L. W. (1967). Enzymatic defenses of certain snails against metal ions. *Veliger*, **10**, 188–91.

Wood, J. H. (1973). Metabolic cycles for toxic elements in aqueous systems. *Revue Internationale d'Océanographie Médicale*, **31/2**, 7–16.

Yonge, C. M. (1926). Structure and physiology of the organs of feeding and digestion in *Ostrea edulis*. *Journal of the Marine Biological Association of the United Kingdom*, **14**, 295–386.

Young, M. L. (1974). The transfer of ^{65}Zn and ^{59}Fe along two marine food chains. PhD thesis, University of East Anglia.

ÅKE LARSSON, BENGT-ERIK BENGTSSON
& OLOF SVANBERG

Some haematological and biochemical effects of cadmium on fish

Most information about the effects of environmental pollutants on aquatic animals has been obtained from mortality studies. Often very little is known about damage to different internal organs or about disturbed physiological and biochemical processes within an organism following exposure to environmental poisons. Consequently, knowledge about the mode of action of toxicants and causes of death in poisoned aquatic animals is often lacking. A better understanding of these mechanisms is necessary if we want to predict the potential harmfulness of various chemicals to the environment.

At our laboratory various means of investigating sublethal effects of pollutants on aquatic animals are used or are under development. The different approaches include study of effects on growth, changes in fecundity or brood survival, changes in swimming activity, and physiological, biochemical and histopathological alterations. So far, most of the studies are performed on fish because of their larger size and our more detailed knowledge about normal functioning and behaviour of these animals compared to aquatic invertebrates.

This paper describes how a clinical toxicological approach has been used to determine haematological and biochemical effects on fish exposed for 15 days to subacute concentrations of cadmium in the water. The aim of the investigation has been to get information about internal disturbances before the fish show apparent external signs of poisoning and to increase understanding of the mode of action of the toxicant. The data presented are part of an investigation on cadmium toxicity in three species of teleost living in the brackish water of the Baltic (Larsson *et al.*, unpublished data). Only the results from one of the fish species tested, the flounder (*Pleuronectes flesus*), will be given here.

Experimental

Flounders with an average weight of 162 g were caught in October in a coastal area of the Baltic (salinity 7‰) near the laboratory. After transport to the laboratory the fish were kept outdoors for 7 weeks in large PVC tanks with a continuous flow of brackish water (salinity 7‰; temperature 7–10 °C). During this period they were fed daily with commercial salmon parr food and live small fish (gobies).

One week before the test started, the fish were transferred to the test aquaria containing 50 l of brackish water. A continuous flow of brackish water from the Baltic was provided by a water flow system described earlier (Bengtsson *et al.*, 1975). The flow rate in each aquarium was 24 l h^{-1}, corresponding to a 90 % test water replacement within 6.5–7 h.

The test water, analysed twice a week according to procedures described by Bengtsson *et al.* (1975), had the following physical and chemical characteristics (expressed as mean ± S.D. of six analyses): temperature (°C), 7.1 ± 0.2; pH, 7.78 ± 0.05; dissolved oxygen (mg l^{-1}), 10.7 ± 0.1; salinity (‰), 7.0 ± 0.1.

Stock solutions of the test substance, $CdCl_2.2\frac{1}{2}H_2O$, were dosed by a peristaltic pump. The concentrations of Cd^{2+} were 0.1, 1 and 10 mg l^{-1}. In each aquarium four flounders were exposed to either test medium or to clean brackish water (controls). Eight fishes per concentration were tested in all. The flounders were not fed during the experiment.

After 15 days of exposure the fish were removed from the water and immediately stunned with a blow on the head. The standardised sampling procedures for blood, liver and muscle tissue as well as the routine clinical methods used in this experiment have been described in detail previously (Larsson, 1972).

Results and discussion

During the time of exposure observations were made in order to detect external signs of poisoning on the experimental animals. No visible symptoms of toxic reaction were seen in the fish exposed to 0.1 and 1 mg l^{-1} of Cd^{2+}. However, after 9–10 days of exposure, the flounders in 10 mg Cd^{2+} l^{-1} exhibited some symptoms of poisoning, such as a slightly excitable condition and occasional unco-ordinated swimming movements. On day 13, one individual died in this highest concentration, the only case of mortality observed during this experiment.

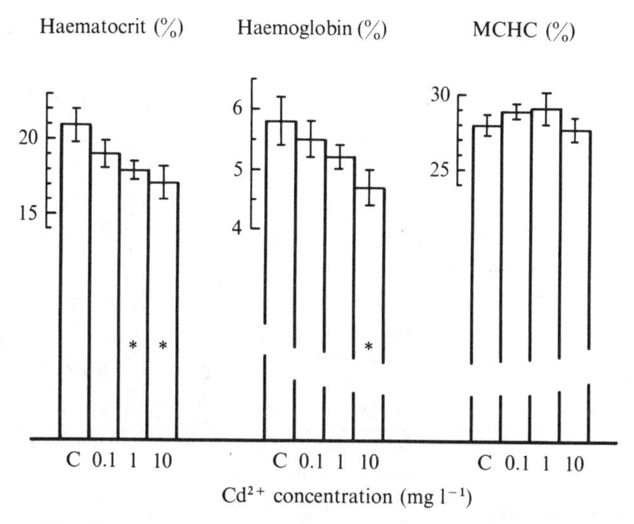

Fig. 1. Haematological characteristics in flounders exposed for 15 days to various concentrations of cadmium in the water. MCHC, mean corpuscular haemoglobin concentration. Each column represents the mean value and vertical bars indicate the standard error of the mean. An asterisk indicates significant difference from the control group (C) at the 0.05 level.

Haematological parameters

Fifteen days of exposure to cadmium caused a dose-dependent depression of the haematocrit and haemoglobin values, most pronounced at the highest concentration (Fig. 1). In 0.1 and 1.0 mg l^{-1} of Cd^{2+} the reduction of haematocrit (9.1 and 14.4 % respectively) was somewhat greater than the haemoglobin decrease (5.2 and 10.3 % respectively). Such a discrepancy suggests that the red blood cells were slightly shrunken in the fish exposed to the two lower concentrations. A small shrinkage of the red blood cells is consistent with the slight increase in the major plasma electrolytes, sodium and chloride, observed in the flounders exposed to the corresponding cadmium concentrations (see below). In the highest cadmium concentration the reduction of haematocrit and haemoglobin was of the same magnitude (18.2 and 19.0 % respectively). The mean corpuscular haemoglobin concentration (MCHC) was not significantly affected by cadmium exposure (Fig. 1).

Collectively the depressions in haematocrit and haemoglobin values indicate that the cadmium-exposed fish were anaemic. Anaemia is often due to an increase of plasma volume caused by a disturbed water balance. Such an explanation is, however, not probable in this case as the plasma total protein levels were not affected by cadmium (Fig. 2). A more plausible explanation is that the anaemia is due to a decreased rate of production of and/or to an increased loss or destruction of red blood cells. An anaemic effect of cadmium is well-documented in higher vertebrates. Swedish and Japanese workers chronically exposed to cadmium dust have lower haemoglobin levels and lower numbers of red blood cells than are regarded as normal (Nilsson, 1970). A cadmium-induced blood anaemia has also been observed in experiments with birds (Freeland & Cousins, 1973) and mammals (Friberg, Piscator & Nordberg, 1971). The exact mechanism for the cadmium-induced anaemia is not fully understood, but seems to be associated, at least partly, with disturbed iron metabolism caused by decreased intestinal absorption (Freeland & Cousins, 1973; Richardson, Fox & Fry, 1974). Worth noting in connection with haematological effects of cadmium is the observation by Gardner & Yevich (1970) that another euryhaline teleost, *Fundulus heteroclitus*, acutely exposed to 50 mg Cd^{2+} l^{-1}, showed similar disturbances in the white blood cell picture (so-called eosinophilia) as those known for cadmium-affected mammals (Nilsson, 1970).

Plasma electrolytes

Sodium and chloride. The major electrolytes in blood plasma, sodium and chloride, showed elevated levels in the two lowest cadmium concentrations (Fig. 2). The plasma sodium content in flounders exposed to 1 mg Cd^{2+} l^{-1} increased by 3.2 % over that of the control fishes, while plasma chloride increased by 10.7 % in 0.1 and by 9.1 % in 1 mg Cd^{2+} l^{-1}. For the time being we have no satisfactory explanation for these elevated plasma sodium and chloride levels and the resultant elevated serum osmolality which occurred only in the two lowest cadmium concentrations. The flounders in the present study were in hypotonic brackish water (Na^+, 95 mM; Cl^-, 109 mM). Pathological changes in the gills, which have been observed previously in cadmium-exposed fish (Gardner & Yevich, 1970; Bilinski & Jonas, 1973) cannot be excluded, but in such a case an osmoregulatory failure would rather have led to a decreased active uptake of monovalent ions from, or simply a loss of these ions to, the hypotonic external media, resulting in decreased levels of the major plasma electrolytes. Thus, the effect may either be due to a stimulated active uptake of sodium and chloride ions from the external media or to a loss of these ions from other tissues to the blood. The latter explanation is in line with the effect observed by Eisler & Edmunds (1966) that another environmental poison, endrin, caused a fall in liver sodium content and a concomitant increase in serum sodium in a marine teleost. An increased blood serum osmolality has also been observed in normally hyperosmotic estuarine crabs (*Carcinus maenas*) exposed to cadmium (Thurberg, Dawson & Collier, 1973). The described cadmium effects on plasma sodium and

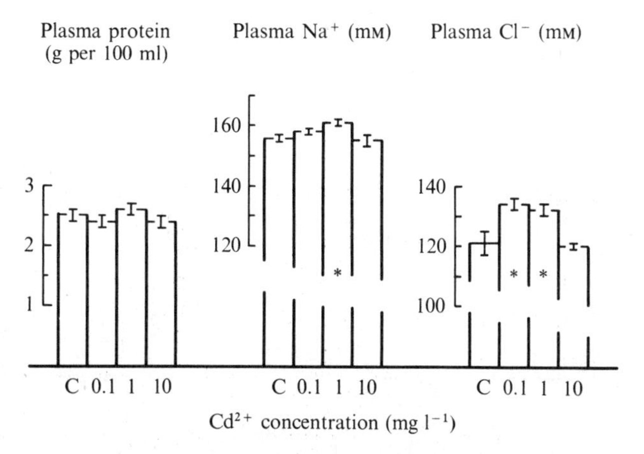

Fig. 2. Total protein, sodium (Na^+) and chloride (Cl^-) in blood plasma from flounders exposed for 15 days to various concentrations of cadmium in the water. Symbols as in Fig. 1.

chloride in the flounders may also have a connection with the observed changes in levels of other ions discussed below.

Potassium, calcium, magnesium and inorganic phosphate. The effect of cadmium exposure on plasma ions other than sodium and chloride was more pronounced (Fig. 3). The potassium content was reduced by 13.2 % in the lowest and by 18.4 % in the two highest cadmium concentrations. Calcium also showed a successive decrease (from 13.8 % to 27.6 % below control levels) from lower to higher cadmium concentration. On the other hand, the plasma content of magnesium ions was remarkably elevated in the cadmium-exposed flounders. The increase, if expressed as a percentage, of this ion was between 114 and 168 % above the control level. Also the plasma content of inorganic phosphate was significantly elevated (17.9 %) in the two highest cadmium concentrations.

The kidney is the main cadmium-sequestering organ and therefore disturbed renal function is a common symptom of acute and chronic cadmium poisoning in mammals (Friberg *et al.*, 1971). This kidney disease is often associated with increased levels of proteins, amino acids, glucose and calcium in the urine. Such a reabsorption defect is due to damage (atrophy and distortion) of the renal tubules. This kidney effect is a typical syndrome in chronic industrial cadmium poisoning as well as in the painful cadmium-caused Itai-Itai disease (ouch-ouch disease) known from certain areas in northern Japan (reviewed by Nordberg, 1974). Furthermore, the increased excretion of calcium through the urine together with an impaired intestinal calcium

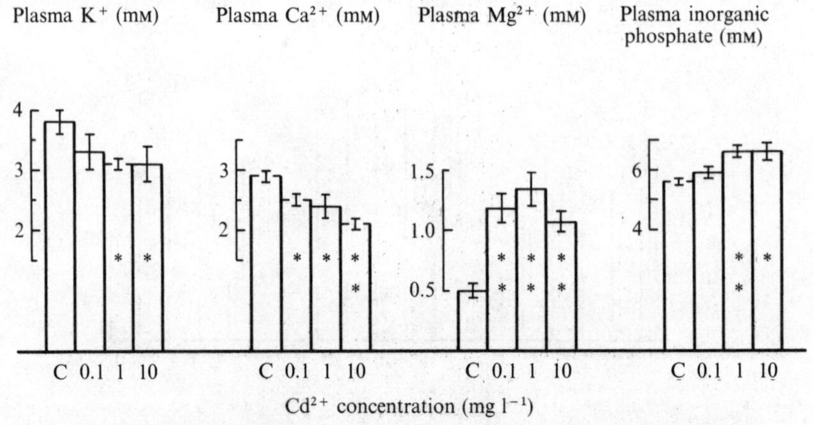

Fig. 3. Potassium, calcium, magnesium and inorganic phosphate in blood plasma from flounders exposed for 15 days to various concentrations of cadmium in the water. One asterisk indicates significant difference from the control group (C) at the 0.05 level, two asterisks at the 0.005 level. Other symbols as in Fig. 1.

absorption mechanism are considered to be the reason for the decalcification of the skeleton with accompanying skeletal deformations and fractures typical of Itai-Itai disease (Nilsson, 1970; Feldman & Cousins, 1973).

Also in fish, tissue damage has been reported as a consequence of acute cadmium poisoning (Gardner & Yevich, 1970). Thus, histological examinations of a cadmium-exposed euryhaline teleost (*Fundulus heteroclitus*) revealed changes in the kidney and the intestinal tract resembling the pathological changes that have been demonstrated both clinically and experimentally in higher vertebrates including man. It might be possible that the observed alterations in the plasma ion contents of the cadmium-exposed flounders in the present study are related to such damage to these important ion-regulating tissues. Hypocalcaemia as well as an increased plasma phosphate content, which were observed in cadmium-poisoned flounders, have also been demonstrated experimentally in cadmium-treated mammals (Kennedy, 1966; Itokawa *et al.*, 1974) and have been interpreted as being due mainly to kidney dysfunction. Also the increased plasma magnesium level could be an effect of kidney damage. The fish in this investigation lived in brackish water with a magnesium ion content of about 9 mM, which is approximately eighteen times higher than the extracellular level. Accordingly, an active renal excretion of this divalent ion must normally occur (Prosser, 1973). The pronounced cadmium-induced elevation of plasma magnesium in the flounders might therefore be due to an impaired ability of the fish to actively excrete the excess of this ion through the kidney. One interesting parallel is that in man certain renal failures are often associated with an elevated serum magnesium concentration (Harper, 1971).

The observed effects of cadmium on the plasma levels of inorganic ions, especially those on potassium and calcium (Fig. 3), may seriously affect important physiological functions of the animal. Potassium is the principal intracellular cation and is intimately involved in nerve and muscle functions. Furthermore, this ion is an important constituent of the extracellular fluid because of its influences on muscle activity. In mammals a lowered serum potassium level (hypokalaemia) often causes muscle weakness, irritability and paralysis, as well as disturbed heart activity (Harper, 1971). Calcium is a general regulator of permeability of cell membranes to water and ions; low calcium usually increases permeability and high calcium decreases it (Prosser, 1973). In addition, calcium is essential for maintenance of the membrane potential and development of the action potential in muscles and nerves (Harper, 1971). In mammals a decrease in serum calcium (hypocalcaemia) is known to cause neuromuscular hyperexcitability and also so-called hypocalcaemic tetanus, i.e. the muscle is repeatedly stimulated and the individual responses fuse into one continuous contraction (Ganong, 1971; Harper, 1971).

Such well-known neuromuscular effects of hypokalaemia and hypocalcaemia together with a decalcification of the skeletal bone might be the explanation for the high incidence of vertebral damage previously observed at our laboratory in minnows (*Phoxinus phoxinus*) exposed to cadmium (Bengtsson *et al.*, 1975). In that investigation vertebral fractures developed in the fish within 70 days when exposed to cadmium in concentrations between 7.5 $\mu g\ l^{-1}$ and 4.7 $mg\ l^{-1}$. The first noticeable symptom in these fish was a highly excitable condition, followed after some weeks of exposure by morphological aberrations (hump-backed condition). X-ray examination of the spinal column revealed fractured vertebrae, especially in the caudal region, in many of the surviving fishes. The fractures might be related to frequently observed vigorous spasms or tetanic muscle contractions. The resulting spinal deformation seriously interfered with the swimming capability of the fish. Vertebral damage was most frequent in fish exposed for 70 days in 0.96 mg $Cd^{2+}\ l^{-1}$, but fractures occurred even at concentrations as low as 7.5 μg $Cd^{2+}\ l^{-1}$. More studies are necessary to confirm the possible relationship between the cadmium-induced vertebral damage in minnows and the ionic imbalances observed in the present flounder experiment.

Liver size

One of the most prominent effects on the flounders after cadmium exposure was the great reduction of the liver size. The liver-somatic index, LSI (= liver

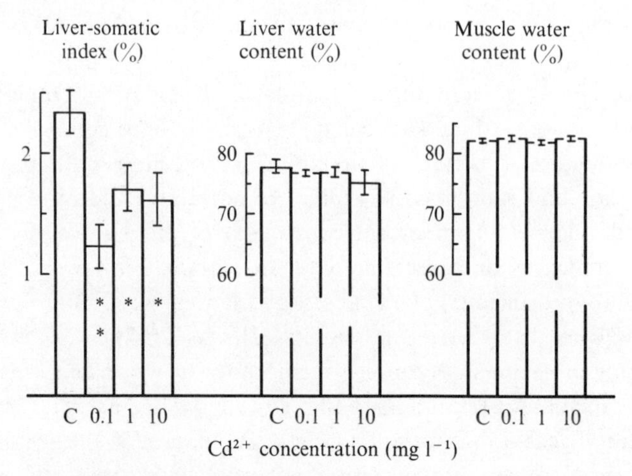

Fig. 4. Liver-somatic index (LSI) and water content in liver and muscle tissue in flounders exposed for 15 days to various concentrations of cadmium in the water. Symbols as in Fig. 1.

weight × 100/body weight), showed a significant decrease (between 27 and 43 %) in all cadmium concentrations (Fig. 4). This effect was not due to a drop in the water content of the liver, as this parameter was unchanged (Fig. 4). The reason for the reduction in liver size may perhaps be better elucidated when planned complementary histological and biochemical examinations have been done. In fish (Tafanelli, 1972) as well as in higher vertebrates (Nordberg, 1974) the liver, next to the kidney, is one of the major storage organs for cadmium. It has also been shown in mammalian experiments that cadmium induces various changes in the liver, from the altered activity of certain liver enzymes to severe liver cirrhosis (reviewed by Nilsson, 1970). However, as suggested by Nordberg (1974), cadmium-induced effects on the liver have probably been overlooked in the past. In fish cadmium has previously been shown to alter the activity of certain liver enzymes (Jackim, Hamlin & Sonis, 1970).

Carbohydrate metabolism

One common clinical finding in cadmium-poisoned mammalian species is an elevated blood sugar level (hyperglycaemia) and an increased urinary glucose level (glucosuria), the latter caused by impaired renal reabsorption. During chronic administration of cadmium the hyperglycaemic response is not observable due to the high renal glucose excretion (Ghafghazi & Mennear, 1973). One purpose of the present investigation was to ascertain whether the same disturbances in carbohydrate metabolism could be detected in cadmium-

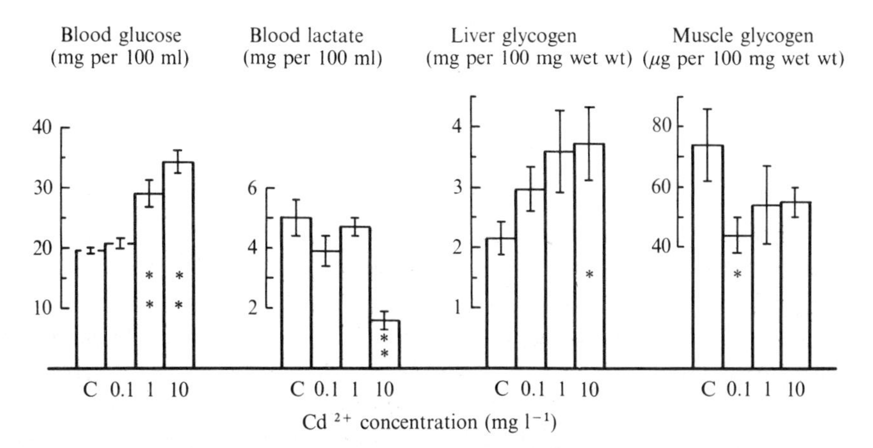

Fig. 5. Blood glucose, blood lactate, liver glycogen and muscle glycogen in flounders exposed for 15 days to various concentrations of cadmium in the water. Symbols as in Fig. 1.

poisoned fish. The observed elevation of blood glucose increased with exposure concentration with a maximum change (75 %) at 10 mg Cd^{2+} l^{-1} (Fig. 5). Increased blood glucose level is mainly due to enhanced breakdown of liver glycogen (glycogenolysis). The unchanged or slightly increased liver glycogen content (Fig. 5) suggests a stimulated glycogen synthesis contemporaneous with the increased liver glycogenolysis and subsequent blood glucose elevation. This is consistent with the finding by Singhal *et al.* (1974) that daily cadmium injections for 45 days to rats enhanced the gluconeogenesis, evidenced by significant increases in the activity of rate-limiting enzymes involved in the transformation of non-carbohydrate metabolites to carbohydrates. Thus, it is probable that the observed effect of cadmium on carbohydrate metabolism is mediated by adrenocortical hormones. Another contributing factor might be reduced insulin secretory activity. Such an explanation is supported by the finding by Havu (1969) that cadmium appears to accumulate in the pancreatic islets of fish and selectively damage insulin-producing beta cells. Cadmium is also found to accumulate in appreciable quantities in the mammalian pancreas (Berlin & Ullberg, 1963) and to reduce circulating serum insulin (Ghafghazi & Mennear, 1973). An impaired secretion of insulin agrees well with the observation that muscle glycogen was slightly decreased in cadmium-exposed flounders (Fig. 5). Thus, glucose uptake in muscle and subsequent muscle glycogen formation seemed to be retarded. The significantly diminished blood lactate level in flounders exposed to 10 mg Cd^{2+} l^{-1} cannot be explained at present.

Conclusion

The complex clinical picture noted in cadmium-exposed flounders resembles, in many respects (i.e. blood anaemia, divalent ion disturbances, altered carbohydrate metabolism), cadmium toxicosis in mammals. This complex cadmium syndrome is not surprising. It is well-established that cadmium has a high affinity for sulphydryl groups especially, but also for hydroxyl groups and ligands containing nitrogen (Nilsson, 1970). Thus, binding to such groups in essential enzymes or enzyme systems might affect various basic biochemical and physiological processes and thereby interfere with and adversely disturb central functions of the organism even at very low cadmium concentrations.

In conclusion the present results indicate that subacute levels of cadmium in the water produce dysfunctions of several physiological and biochemical processes in fish. At present it is difficult to evaluate the ecological significance of these relatively short-term alterations. However, the results encourage further work with still lower concentrations of cadmium and with longer test periods. Such sublethal studies are important, as even small physiological disturbances might reduce the chance of the animal being successful in its environment. It must also be emphasised that elimination of aquatic animals by small, insidious physiological or behavioural changes can be regarded as more serious than a massive fish kill, since it is less likely to be observed and corrected.

References

Bengtsson, B.-E., Carlin, C. H., Larsson, Å. & Svanberg, O. (1975). High incidence of vertebral damage in minnows, *Phoxinus phoxinus* L., exposed to cadmium. *Ambio*, **4**, 166–8.

Berlin, M. & Ullberg, S. (1963). The fate of Cd^{109} in the mouse. *Archives of Environmental Health*, **7**, 686–93.

Bilinski, E. & Jonas, R. E. E. (1973). Effects of cadmium and copper on the oxidation of lactate by rainbow trout (*Salmo gairdneri*) gills. *Journal of the Fisheries Research Board of Canada*, **30**, 1553–8.

Eisler, R. & Edmunds, P. H. (1966). Effects of endrin on blood and tissue chemistry of a marine fish. *Transactions of the American Fisheries Society*, **95**, 153–9.

Feldman, S. L. & Cousins, R. J. (1973). Influence of cadmium on the metabolism of 25-hydroxycholecalciferol in chicks. *Nutrition Reports International*, **8**, 251–60.

Freeland, J. H. & Cousins, R. J. (1973). Effect of dietary cadmium on anemia, iron absorption, and cadmium binding protein in the chick. *Nutrition Reports International*, **8**, 337–43.

Friberg, L., Piscator, M. & Nordberg, G. (1971). *Cadmium in the environment*. CRC Press, Cleveland, Ohio.

Ganong, W. F. (1971. *Review of medical physiology*. Lange Medical Publications, Los Altos, California.

Gardner, G. R. & Yevich, P. P. (1970). Histological and hematological responses of an estuarine teleost to cadmium. *Journal of the Fisheries Research Board of Canada*, **27**, 2185–96.

Ghafghazi, T. & Mennear, J. H. (1973). Effects of acute and subacute cadmium administration on carbohydrate metabolism in mice. *Toxicology and Applied Pharmacology*, **26**, 231–40.

Harper, H. A. (1971). *Review of physiological chemistry*. Lange Medical Publications, Los Altos, California.

Havu, N. (1969). Sulfhydryl inhibitors and pancreatic islet tissue. *Acta Endocrinologica*, Supplement, **139**, 1–231.

Itokawa, Y., Abe, T., Tabei, R. & Tanaka, S. (1974). Renal and skeletal lesions in experimental cadmium poisoning. *Archives of Environmental Health*, **28**, 149–54.

Jackim, E., Hamlin, M. & Sonis, S. (1970). Effects of metal poisoning on five liver enzymes in killifish (*Fundulus heteroclitus*). *Journal of the Fisheries Research Board of Canada*, **27**, 383–90.

Kennedy, A. (1966). Hypocalcemia in experimental cadmium poisoning. *British Journal of Industrial Medicine*, **23**, 313–17.

Larsson, Å. (1972). Metabolic studies on eels. A presentation of routine procedures for metabolic studies on fish, and of their use in investigations of alterations in the over-all metabolism of the eel after different treatments. PhD thesis, Department of Zoophysiology, University of Göteborg, Sweden.

Nilsson, R. (1970). *Aspects on the toxicity of cadmium and its compounds: a review. Ecological Research Committee, Bulletin No. 7*. Swedish Natural Science Research Council, Stockholm. 49 pp.

Nordberg, G. F. (1974). Health hazards of environmental cadmium pollution. *Ambio*, **3**, 55–66.

Prosser, C. L. (1973). *Comparative animal physiology*, 3rd edition, pp. 46–106. W. B. Saunders, Philadelphia.

Richardson, M. E., Fox, M. R. S. & Fry, B. E. Jr (1974). Pathological changes produced in Japanese quail by ingestion of cadmium. *Journal of Nutrition*, **104**, 323–38.

Singhal, R. L., Merali, Z., Kacew, S. & Sutherland, D. J. B. (1974). Persistence of cadmium-induced metabolic changes in liver and kidney. *Science*, **183**, 1094–6.

Tafanelli, R. (1972). Toxicity and pathogenic effects of intraperitoneal injections of cadmium chloride to the goldfish. Thesis, Oklahoma State University.

Thurberg, F. P., Dawson, M. A. & Collier, R. S. (1973). Effects of copper and cadmium on osmoregulation and oxygen consumption in two species of estuarine crabs. *Marine Biology*, **23**, 171–5.

R.LLOYD & D.J.SWIFT

Some physiological responses by freshwater fish to low dissolved oxygen, high carbon dioxide, ammonia and phenol with particular reference to water balance

Introduction

Increasing attention has been given during the past decades to the protection of the aquatic environment against pollution, both nationally and internationally; in the freshwater field, the presence or absence of fish has been widely used as a biological indicator of the degree of pollution. Much of the research has been to obtain data on the concentrations or values of selected individual pollutants which are lethal to fish either in the short- or long-term. Such data can provide very necessary information, apart from identifying a boundary limit above which fish are likely to be killed. For example, they can indicate the relation between the effects of environmental factors, such as water hardness, temperature, pH value and dissolved oxygen concentration, and the toxicity of the pollutant. Studies have shown, for example, that the toxicity of heavy metal salts is strongly dependent on the calcium concentration of the water, that ammonia toxicity is related to the pH value and temperature, and that several other poisons appear to be more harmful at low levels of dissolved oxygen, probably because of the influence on the respiratory rate (Lloyd, 1961). Other effects, such as the comparative sensitivity of different stages in the life history or interspecific differences, can also be demonstrated by measuring the lethal response by fish to pollutants.

However, in order to maintain healthy stocks of fish in our inland waters, water quality objectives have to be determined for each pollutant which ensure that the species present are not adversely affected in any way. For this reason attention has been given to sublethal responses of fish to selected pollutants; such responses range from biochemical to behavioural and a critical examination of some of these has been made by Sprague (1971). It suffices to reiterate here that there is no one sublethal test which alone can provide a basis for water quality objectives; life history studies, such as those carried out at the Water Quality Laboratory, Duluth, USA (Mount & Stephan, 1967), probably give more information than any other single test, but they do not completely simulate the normal aquatic environment and are time-consuming to undertake. Studies on growth rates have the advantage of relative simplicity and are important for commercial fisheries, but the general thesis that healthy fish exhibit normal growth rates may not always be correct; for example, survivors of batches of rainbow trout exposed to chronically lethal solutions of zinc for 180 days showed a growth rate similar to the control batch (Ministry of Technology, 1967), whereas with exposure to sublethal cadmium solutions the growth rate increased slightly with increasing concentration, and decreased with exposure to increasing sublethal concentrations of copper (Department of the Environment, 1973). Coho salmon exposed to bleached kraft mill effluent at 25 % of the lethal concentration grew more rapidly than controls in clean water (McLeay & Brown, 1974) when fed with an excess ration. Such studies are complicated by the possibility that a laboratory feeding regime may be less taxing to fish than food-gathering in the natural environment. Many other biochemical or physiological measurements have been made in the hope of establishing suitable criteria, but in general they fail to satisfy one or more of three main ideal requirements; that (*a*) differences detected as a result of an imposed stress are greater than the natural variability of the response, (*b*) a graded response to the applied stressor exists allowing the maximum stressor level giving no effect to be measured; and (*c*) the response measured is not solely part of an adaptive process which maintains homeostasis without significant metabolic cost.

Some difficulties have arisen from a confusion over the term 'stress'; in this paper the definition given by Brett (1958) is used: 'Stress is a state produced by an environmental or other factor which EXTENDS the adaptive responses of an animal beyond the normal range, or which disturbs the normal functioning to such an extent that, in either case, the chances of survival are significantly reduced.' By this definition a fish which alters its internal physiology without significant extra metabolic cost (i.e. within its normal range) to counteract an applied stressor is not stressed, and a change in an adaptive physiological reaction measured under these conditions does not necessarily represent a detrimental effect.

During the past decade research has been carried out at the Salmon and Freshwater Fisheries Laboratory, London, into some of the physiological and biochemical responses of fish to applied stressors, including free carbon dioxide, low dissolved oxygen, ammonia and phenol, all of which are common in polluted waters; these will be considered separately.

General observations

Physiological response to low levels of dissolved oxygen

Discharge of oxidisable organic matter into watercourses, giving rise to a reduction in the dissolved oxygen concentration, is a major cause of pollution, though even in the absence of organic pollution dissolved oxygen levels can fluctuate diurnally in response to the photosynthetic and respiratory activity of aquatic plants and, in heavily weeded enclosed waters, a reduction in dissolved oxygen concentration of up to 50 % of the air saturation value can occur during the night. This can have concomitant effects on the tissue oxygen levels of rainbow trout (Garey & Rahn, 1970). A considerable literature exists on the effects of hypoxia on the survival, growth, activity, avoidance reaction and other responses of fish (critically reviewed by Doudoroff & Shumway, 1970, and EIFAC, 1973) and on respiratory responses (reviewed by Hughes, 1973). In many experiments, fish conditioned to aerated water have been suddenly exposed to hypoxic conditions and the immediate responses then measured. However, such conditions are not encountered normally by fish in the environment; dissolved oxygen levels tend to change slowly both spatially and temporally if there is chronic organic pollution.

It is known that fish can acclimate to low levels of dissolved oxygen. For example, the classical studies of Shepard (1955) showed that brook trout held at moderately low levels of dissolved oxygen were more resistant to very low levels which were lethal to fish previously held in aerated water.

A review of the limited literature on the responses to prolonged hypoxia is given by Randall (1970). It is known, *inter alia*, that haematocrit values tend to increase, but whether this is caused by erythrocyte swelling or to an increased number of red blood cells is uncertain; the relevant data have been reviewed by Doudoroff & Shumway (1970). Using rainbow trout Swift & Lloyd (1974) found that the haematocrit increased significantly from a value of 38 to 50 % within 3 hours of exposure to 3 mg O_2 l^{-1} at 12–16 °C, and that slightly less elevated levels were found at the end of a 24-hour exposure to 4.5 mg O_2 l^{-1}. It was observed that the erythrocyte count of trout had increased considerably at the end of 24-hour exposure to hypoxia but that there was no change in erythrocyte volume. Haemoglobin content of the erythrocytes appeared to be unchanged, but the increased haematocrit led to a greater haemoglobin concentration in the blood.

An elevated haematocrit can be achieved either by haemoconcentration (reduction in blood volume) or by release of stored erythrocytes from the spleen (Hall, 1928; Privolnev, 1954; Ostroumova, 1954; Stevens, 1968). Direct measurements of blood volume of fish present considerable technical difficulties (Smith, 1966) and therefore an indirect method has been used on

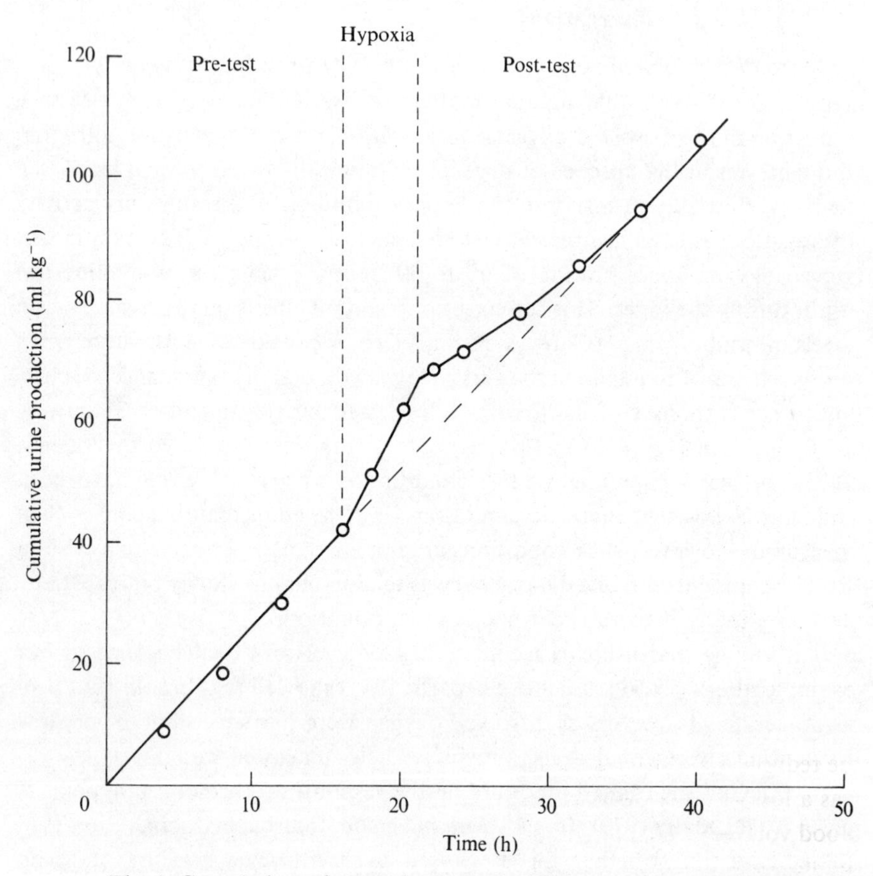

Fig. 1. Cumulative urine production of rainbow trout ($n = 8$) before, during and after a 5-hour exposure to hypoxia (3 mg O_2 l^{-1}). (From Swift & Lloyd, 1974.)

trout in which changes in urine flow rate (UFR) were measured (Swift & Lloyd, 1974). The fish tested were acclimated to the apparatus for 2 days and then, after being fitted with a urinary catheter, were left for 1 day to recover. UFRs were monitored from the second day onwards to obtain a 'pre-hypoxic' rate before the dissolved oxygen concentration was lowered to 3 mg O_2 l^{-1} for up to 5 hours. Data from one experiment (Fig. 1) clearly indicate that the UFR increased during the hypoxia period (values were 2.6 ± 0.6 ml kg^{-1} h^{-1} for the pre-test period and 4.5 ± 0.9 ml kg^{-1} h^{-1} during the hypoxia period). Similar results were obtained for 3–4-hour exposure periods to hypoxia. When the dissolved oxygen levels returned to normal there was a reduction in UFR to 1.9 ml kg^{-1} h^{-1} for about 10 hours, before a gradual return to the normal rate. Fig. 1 shows that the increased UFR during hypoxia was balanced by

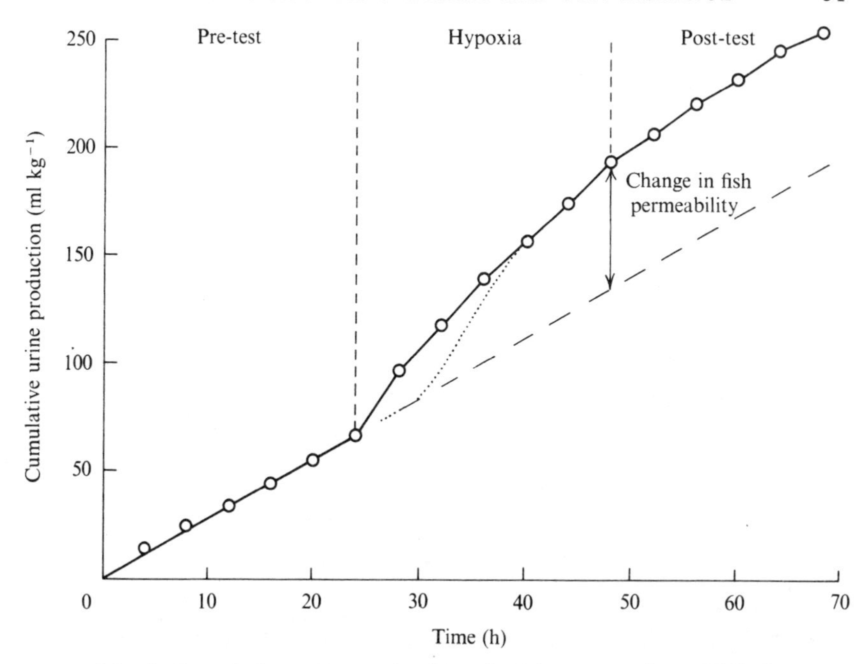

Fig. 2. Cumulative urine production of rainbow trout ($n = 13$) before, during and after 24-hour exposure to 3 mg O_2 l^{-1}. Dotted line indicates postulated change in fish permeability. (From Swift & Lloyd, 1974.)

the reduced UFR during the post-test period, and it can be inferred that there was a loss of body fluid (16 ml kg^{-1}) during the hypoxia period. Assuming a blood volume of 60 ml kg^{-1} (6 %) body weight (Smith, 1966), which may be a slight overestimate but within the range found for salmonid species, the loss of body fluid recorded would account for an increase in haematocrit from 39 to 49 % if it reflected a reduction in blood volume.

During exposure to longer periods of hypoxia (4.5 mg O_2 l^{-1} for 24 hours) the UFR showed an initial increase but no marked subsequent decrease to subnormal levels (Fig. 2). It is possible that an increase in water uptake occurred during hypoxia since the extra urine production greatly exceeded any conceivable reduction in body fluid levels. It is unlikely, from the results of the short-term experiments, that this increase in water uptake occurred within the first 5 hours and there was probably an initial haemo-concentration associated with reduction in blood volume. The return of the blood volume to normal may also have occurred within the 24-hour hypoxic period since no diminution of UFR occurred during the post-test period; however, any fluid retention after the initial 5-hour hypoxic period could have been masked by the increase in water uptake rate. If the change in water uptake rate during this time follows the dotted line in Fig. 2, the blood volume could have returned to normal during the latter part of the hypoxic period; the sustained elevated haematocrit values observed at the end of the 24-hour hypoxic period could then be accounted for only by a release of erythrocytes from the spleen.

At the present time these hypotheses are based on inference rather than on direct measurement and accurate estimates of blood volume during the different stages of hypoxic exposure are required to substantiate them.

Physiological response to high levels of free carbon dioxide

Although free carbon dioxide is not an important aquatic pollutant in its own right, elevated levels can occur as a result of oxidation of organic matter and it is often associated, therefore, with reduced levels of dissolved oxygen. Much of the existing data on the effects of free carbon dioxide on fish are on its influence in raising lethal hypoxic levels (e.g. Alabaster, Herbert & Hemens, 1957) as well as on oxygen consumption (Fry, 1957; Basu, 1959) and respiratory efficiency (van Dam, 1938; Saunders, 1962). However, it has been shown that the reduction in respiratory efficiency by increased free carbon dioxide lasts for only a few hours (Saunders, 1962). Also, the effect of a moderate concentration of free carbon dioxide on the susceptibility of fish to low concentrations of dissolved oxygen disappears if the fish are acclimated to the level of free carbon dioxide before the experiment (McNeil, 1956, quoted in Davis, Foster, Warren & Doudoroff, 1963; Dahlberg, Shumway & Doudoroff, 1968).

During experiments on the resistance of rainbow trout to acids, it was found that lethal pH values were raised when the free carbon dioxide content of the water was increased (Lloyd & Jordan, 1964). Measurements of the pH value and bicarbonate alkalinity of the blood showed that rainbow trout acclimated for about 6 days to 50 mg CO_2 l^{-1} had blood bicarbonate levels of 30.1 mM kg^{-1} and a pH of 7.21, compared with normal values of 12.4 and 7.20 respectively; at ambient pH values of 4.5 these levels could not be sustained and the blood pH fell to 6.65, with a bicarbonate alkalinity of 13.5 mM kg^{-1}. Additional blood analyses of rainbow trout exposed to elevated carbon dioxide levels (35 mg CO_2 l^{-1}) (Fig. 3) showed that serum bicarbonate levels slowly increased from 6.75 to 32.2 mM l^{-1} over a period of 24 hours, with a concomitant reduction in serum chloride from 152.5 to 114.6 mM; serum sodium concentrations remained unchanged (Lloyd & White, 1967). A return of the free carbon dioxide concentrations to normal values was accompanied by a rapid decrease in the serum bicarbonate level and a slower return of serum chloride to normal.

In mammals acid–base control of blood is a renal function. The urine composition of rainbow trout exposed to high levels of free carbon dioxide

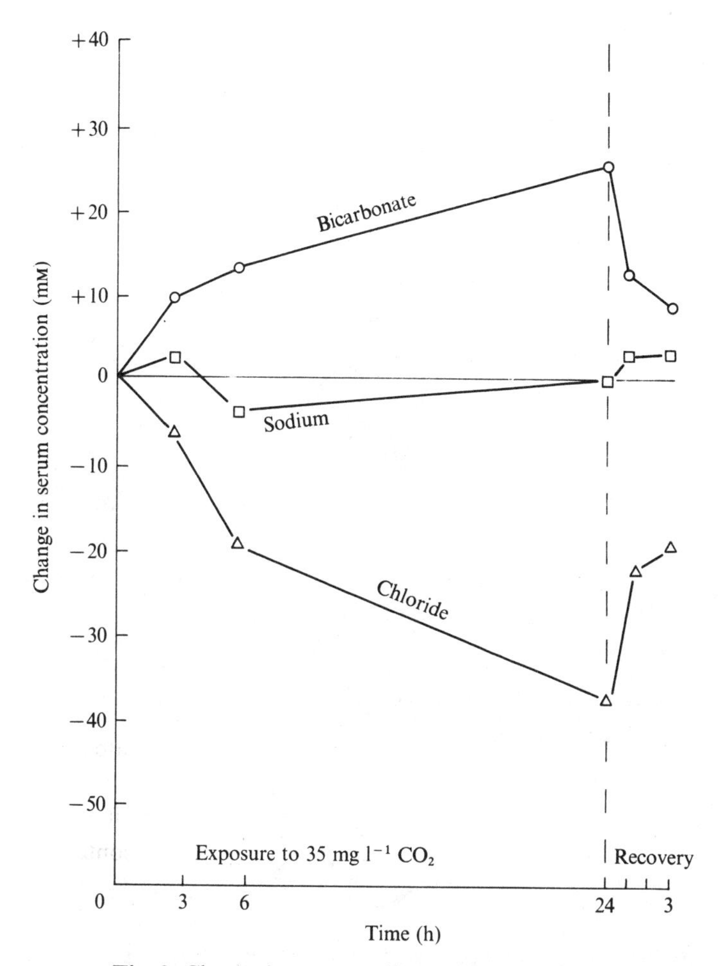

Fig. 3. Change in serum sodium, chloride and bicarbonate concentration of rainbow trout exposed to 35 mg l^{-1} free carbon dioxide. Control values (mM ± S.D.): sodium, 170 ± 5.7; chloride, 152.5 ± 6.8; and bicarbonate, 6.75 ± 2.8. (After Lloyd & White, 1967.)

was therefore examined. In three experiments fish were exposed to normal aerated water for 2 days, followed by a 2-day period during which the free carbon dioxide content was increased to 21 mg l^{-1} with a further 1-day period in normal aerated water. The fish were fitted with a urinary catheter and analyses were made on the urine collected during consecutive 4-hour periods. Of the fish tested one was very active, one moderately active and one inactive during the period of the experiment. Data from one experiment are shown in Fig. 4; because of the variability of the results the data from all three tests could not be combined, although the general pattern of fluctuations was similar.

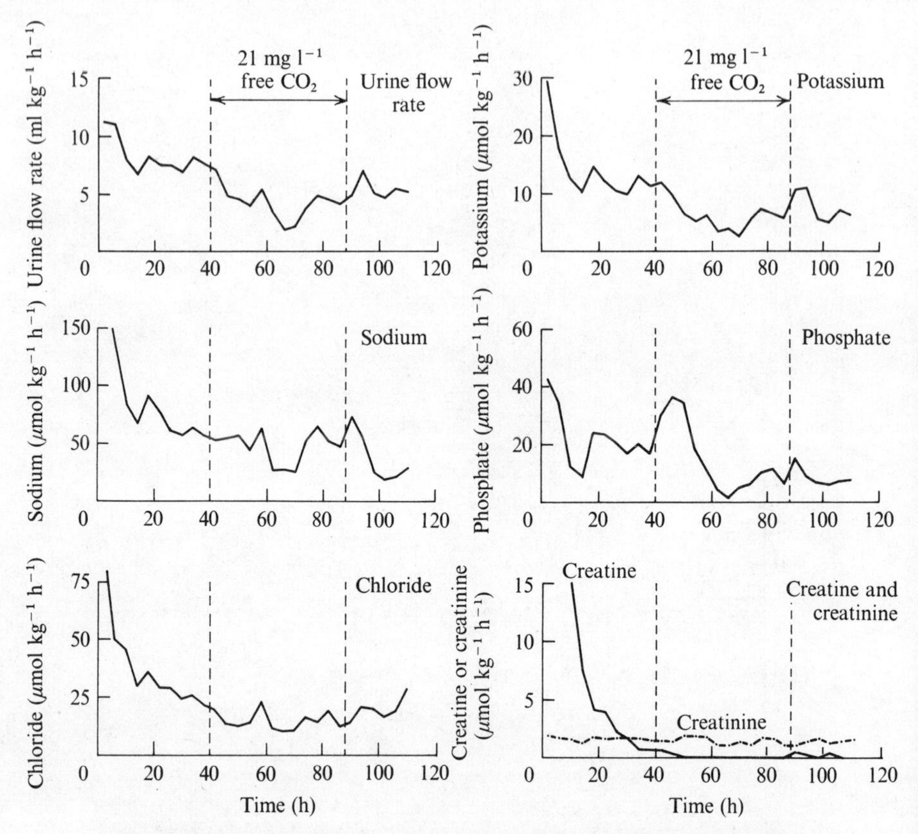

Fig. 4. Values for urine flow rate and excretion of sodium, potassium, phosphate, chloride, creatine and creatinine of a rainbow trout before, during and after 48-hour exposure to 21 $mg\,l^{-1}$ free carbon dioxide.

The very active fish had a high UFR at the start of the test; excretion rates of all the substances measured, except ammonia, were also high but gradually fell to a stable level towards the end of the 2-day pre-test period (Fig. 4). Initial high rates of sodium and chloride loss in the urine were probably correlated with the inability of the kidney to achieve maximal reabsorption of these ions at high UFRs. The concentration of potassium in the urine was fairly constant (range 1.0–2.7 mM) and thus the excretion rate of this ion was directly related to urine flow. Creatinine excretion varied only slightly but initial high concentrations of creatine and phosphate in the urine probably resulted from the high activity of this fish; high phosphate levels have been recorded by Hunn & Wilford (1970). The initial urine flow of the moderately active fish was 8.8 $ml\,kg^{-1}\,h^{-1}$, dropping to an average value of 3.7 $ml\,kg^{-1}\,h^{-1}$ after 4 hours. Excretion rates of creatine and phosphate were initially high (3.0 $\mu mol\,kg^{-1}\,h^{-1}$ and 26 $\mu mol\,kg^{-1}\,h^{-1}$ respectively), falling to a lower level after 16 hours. Rates of excretion of the other ions measured were high at the start, but became normal as the urine flow rate returned to normal. There was no diuretic response by the least active fish and the urine composition remained at a steady level; creatine and phosphate excretion rates were very low even at the start of the test.

When the level of free carbon dioxide in the water was raised to 21 mg l^{-1} the urine flow rates from both active and inactive fish decreased but there was no change in the flow rate from the moderately active fish. The only changes in the urine composition common to all three fish were increases in the non-phosphate alkalinity, the phosphate concentration (which was unaccompanied by an increased creatine output), and the pH value, which fell to about 7.0 in the first 8 hours, but then rose to above 8.0 for the remainder of the period. Reversion to normal levels of free carbon dioxide resulted in an increased urine flow in all three fish for a short period, followed by normal flow rates; non-phosphate alkalinity and pH values returned to normal and there was no significant change in the excretion rates of the other substances measured.

The function of increased phosphate excretion levels at the onset of high ambient carbon dioxide levels remains unclear; no increase in serum phosphate was found by Lloyd & White (1967) under similar experimental conditions. However, Hunn (1969) found that phosphate levels in rainbow trout urine increased after hypoxic stress, and Wood & Johansen (1972) found a reduced ATP content in eel blood during hypoxia associated with an increase in oxygen affinity of the haemoglobin. Eddy & Morgan (1969) showed the Bohr effect in rainbow trout acclimated to high levels of free carbon dioxide was reduced by comparison with that of fish in normal conditions. It is possible, therefore, that the increased phosphate excretion in the urine is associated with changes in the oxygen affinity of the blood.

These three experiments indicate that the ionic changes in urine composition which follow an increase in ambient free carbon dioxide concentration are quantitatively of a very minor order and probably reflect only the changes occurring in the fish blood rather than indicating that the kidney plays any major role in adaptation to conditions of high carbon dioxide.

Further experiments utilised a divided chamber similar to that designed by Post, Shanks & Smith (1965). Essentially the head of the fish, including the gills, was fitted into an enclosed chamber through which aerated water was passed. For a period of one hour in every four the flow of water was stopped and samples for sodium, chloride and ammonia analysis were taken automatically at the beginning and end of the period. The head chamber was aerated during the period of no flow. The fish were allowed to acclimate to the experimental conditions for 2 days, to enable base-line values for ionic fluxes to be obtained, then exposed to 44 mg CO_2 l^{-1} for 1 day followed by

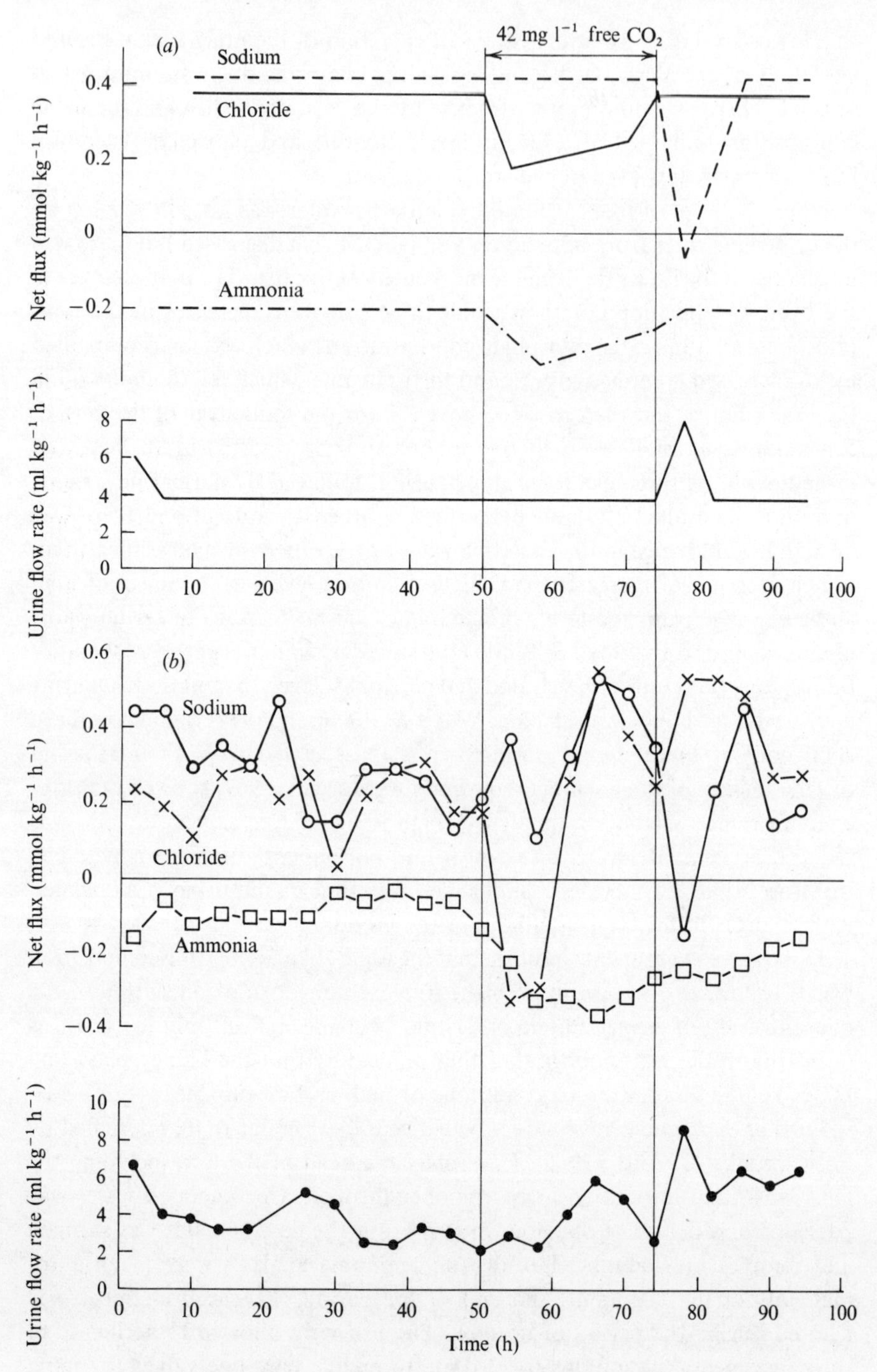

Fig. 5. Values for net sodium, chloride and ammonia fluxes in the head region and urine flow rates of rainbow trout, before, during and after exposure to 42 mg l^{-1} free carbon dioxide. (*a*) General pattern of response; (*b*) data from a single experiment.

Table 1. *Average values for net sodium and chloride uptake and ammonia loss by the head region of rainbow trout before* (2 *days*), *during* (1 *day*) *and after* (1 *day*) *exposure to* 44 *mg* l^{-1} *free carbon dioxide*

Experiment	Period	Sodium (mmol $kg^{-1}h^{-1}$)	Chloride (mmol $kg^{-1}h^{-1}$)	Ammonia (mmol $kg^{-1}h^{-1}$)
1	Pre-test	0.53	0.53	0.21
	Test	0.53	0.26	0.44
	Post-test	0.26	0.19	0.32
2	Pre-test	0.12	0.20	0.18
	Test	0.41	0.15	0.37
	Post-test	0.43	0.60	0.22
3	Pre-Test	0.81	0.58	0.35
	Test	0.65	0.34	0.39
	Post-test	0.28	0.45	0.28
4	Pre-test	0.47	0.40	0.17
	Test	0.43	0.23	0.33
	Post-test	0.27	0.54	0.14
5	Pre-test	0.29	0.22	0.09
	Test	0.36	0.14	0.30
	Post-test	0.18	0.43	0.21

Experimental temperature, 19 ± 1 °C.

a 1-day recovery period. Values obtained for the net sodium and chloride fluxes in the head region showed considerable variations, both between fish and throughout an experiment; this would be expected if the net uptake of these ions by the head region balanced their loss through the posterior integument, and that this loss was increased by integumental damage or activity which may have varied during the test. In general, however, the results conformed to the diagrammatic representation shown in Fig. 5(*a*), and the data from a single experiment are shown in Fig. 5(*b*). Average values for net sodium, chloride and ammonia fluxes for the three test phases for five experiments are shown in Table 1.

Fig. 5(*a*) indicates that an increase in the free carbon dioxide content of the water was followed by a decrease in the net chloride uptake (in the experiment shown in Fig. 5(*b*) there was a net chloride loss) and this was accompanied by an increase in the excretion rate of ammonia by the gills. These changes in ionic fluxes coincided with an increase in the blood serum bicarbonate level and the decrease in blood serum chloride content shown in Fig. 3. When the level of free carbon dioxide in the water returned to normal the net chloride uptake remained unchanged but during the first 4 hours the net sodium uptake by the gills was from 0.19 to 0.67 mmol kg^{-1} h^{-1} less than the previous values

Table 2. *Change in the net sodium flux (mmol kg^{-1} h^{-1}) in the head region of rainbow trout on reduction of the free carbon dioxide content from* 44.0 *to* 1.5 *mg l*$^{-1}$

	Experiment 1	2	3	4	5	6	7	8	Av.
2h before reduction in CO_2	0.34	0.35	0.62	0.43	0.35	0.74	0.14	0.29	0.41
2h after reduction in CO_2	0.13	−0.30	−0.01	−0.24	−0.15	0.15	−0.05	−0.05	−0.06
Difference	0.21	0.65	0.63	0.67	0.50	0.59	0.19	0.34	0.47

(Table 2). This decreased sodium uptake was accompanied by a temporary increase in the rate of urine flow and a subsequent return to normal of the rate of ammonia excretion. However, during this recovery period the serum sodium levels remained constant (Fig. 3) and it seems likely that as bicarbonate ions are rapidly lost from the gills together with sodium ions, an elevated UFR accompanied by haemoconcentration may also take place to maintain the osmolarity of the serum. Although it has been shown that chloride ions are taken up by the gills in exchange for bicarbonate ions (Maetz & Garcia Romeu, 1964), the net uptake of chloride ions during the recovery period does not increase above the normal value (Fig. 5).

The role of increased ammonia excretion during the period of high ambient free carbon dioxide is unknown, but it may reflect an increase in the basic metabolic rate and associated protein catabolism rather than an essential process in ionic regulation. It is clear, however, that regulation of the blood alkalinity in rainbow trout is achieved by the gills and the kidney only plays a part in maintaining serum osmolarity; indeed, in the post-test period when sodium ions were being lost from the gills the kidney still continued to absorb sodium at the normal rate (Fig. 2).

Physiological response to elevated levels of ambient ammonia

Ammonia is a common pollutant in rivers and lakes, often being associated with organic discharges but sometimes appearing in industrial effluents. The literature on the effect of ammonia on fish has been critically reviewed by EIFAC (1970). Although there have been many studies on the direct lethal effect of ammonia, and the effects of pH, temperature, dissolved oxygen and free carbon dioxide on its toxicity, little is known of its sublethal effects. Histopathological studies indicate that gill epithelium hyperplasia can occur in chinook salmon fry (Burrows, 1964) and other external and internal tissues of carp have been found to be extensively damaged by concentrations which were not acutely lethal (Flis, 1968).

The possibility of ammonia increasing the water uptake by fish has been investigated by Lloyd & Orr (1969). Rainbow trout were fitted with a urinary catheter and placed in an apparatus which allowed the UFR to be monitored continuously by means of a drop counter and chart recorder. After a 1-day period of acclimation to the apparatus the UFR was monitored to obtain a base-line reading; ammonia was then added to the water for a 24-hour period after which the fish were exposed to ammonia-free water for a further day. In the first series of tests, batches of rainbow trout were exposed to 3.75, 7.5, 12, 15 and 20 mg NH_3 l^{-1} as N. Although ten fish were used in each experiment not every fish produced a continuous record of urine flow since some catheters became blocked and others were dislodged. The cumulative average urine production for the duration of the test is shown in Fig. 6. The pre-test values are the mean of all the fish used in the five tests since the mean pre-test values for each individual test differed only slightly from each other. It can be seen that the increase in UFR is a function of the ambient ammonia concentration; the highest concentration was lethal and all the fish died during the exposure period. At 15 mg l^{-1}, two out of the six fish died. Under the conditions of pH and temperature of the test, the lethal threshold concentration of ammonia was 17 mg l^{-1}.

The increased UFR was not sustained during the exposure period and in the lower ammonia concentrations the rate had decreased considerably at the end of 24 hours, although it was still above the normal level. There was no reduction in UFR below pre-test rates on return to clean water and, from the amount of extra urine excreted, it was clear that the uptake of water by the fish had increased considerably with exposure to ammonia. It was apparent that the onset of increased UFR was not immediate on elevation of the ambient ammonia concentration but that a time lag occurred which was positively correlated with the concentration of ammonia present.

It is clear, therefore, that the change in UFR is a graded response to an increase in ambient ammonia concentration. The minimum level at which ammonia exerts an effect was determined from the relation between ammonia concentration and the mean and maximum UFR (Fig. 7); a straight line has been fitted by eye to the experimental points and extrapolated to the pre-test mean UFR. It was concluded from these experiments that 12 % of a lethal threshold ammonia concentration has no effect on the water uptake by rainbow trout and, in the absence of other quantitative sublethal data, this value was used in the preparation of water quality objectives for ammonia by EIFAC (1970).

The underlying mechanism by which ammonia increases the water uptake

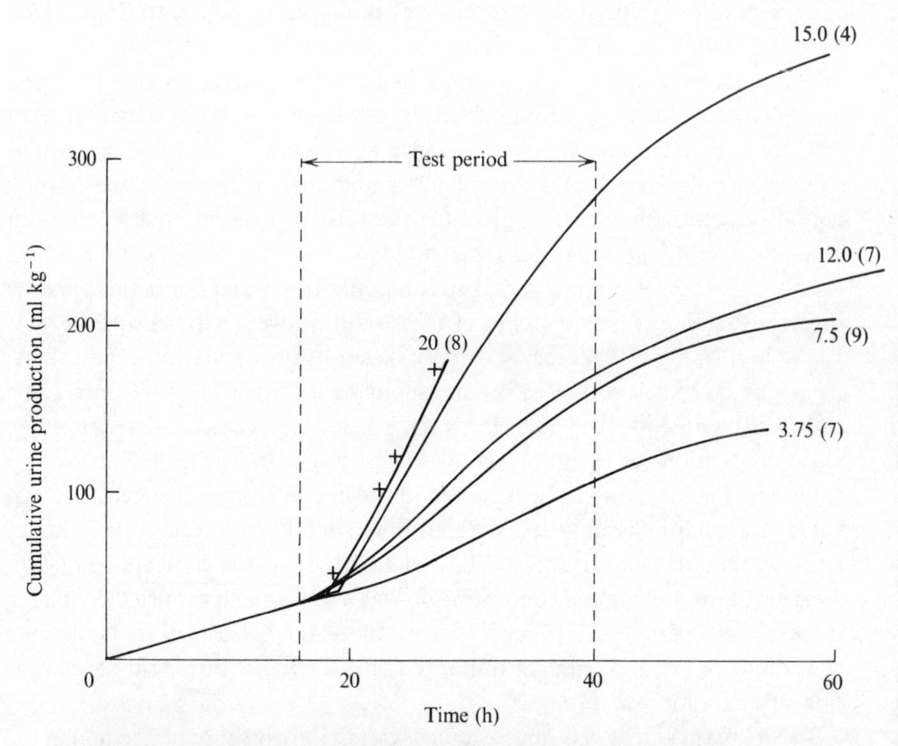

Fig. 6. Effect of concentration of ammonia on urine production of rainbow trout. Ammonia concentrations as mg l^{-1} N are shown against each curve and the number of fish in brackets. Times of death of individual fish are shown as crosses against the curve for 20 mg l^{-1} ammonia as N. Temperature, 10.5 °C; pH, 8.10. (From Lloyd & Orr, 1969.)

by fish is not known. However, the effect is unlikely to be immediate since successive exposures of rainbow trout to ammonia for 2 hours (the time lag for the concentration used), with a 1-hour interval in clean water, did not lead to a subsequent significant increase in UFR. This suggests that the level of ammonia in the tissues did not reach a critical level within the exposure period and that the fish was fully recovered after the 1-hour interval in clean water.

The progressive reduction in UFR after about a 12-hour exposure to ammonia (Fig. 6) indicated that some acclimation was taking place, possibly by an increased rate of ammonia detoxification. Some experiments were carried out in which rainbow trout were exposed to 0.22 mg l^{-1} un-ionised ammonia as N for 24 hours during which time the UFR increased from 2.8 to 6.4 ml kg^{-1} h^{-1}. The ammonia concentration was then doubled and, during the following 15 hours, the UFR increased to 8.0 ml kg^{-1} h^{-1}.

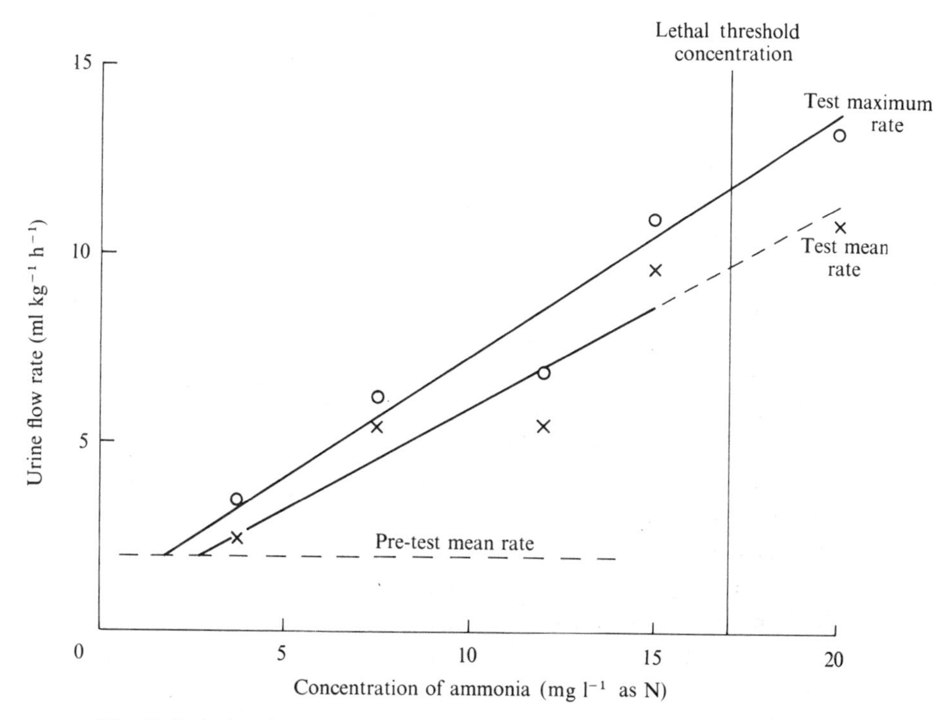

Fig. 7. Relation between maximum and mean flow rates of rainbow trout and the concentration of ammonia. Temperature, 10.5 °C; pH, 8.10. (From Lloyd & Orr, 1969.)

Rainbow trout exposed directly to 0.44 mg l^{-1} un-ionised ammonia all died within 3 hours, their UFR rising from 1.75 to 11.0 ml kg^{-1} h^{-1} during this period. Other tests indicated that the acclimation achieved at the end of a 24-hour exposure was maintained for 24 hours after return to clean water, but was lost within 3 days. Supporting evidence for acclimation by fish to ammonia solutions, based on the lethal response, has been reviewed by EIFAC (1970). It is possible that acclimation depends on the increased synthesis of glutamic acid which is then converted to glutamine in the presence of ammonia; Levi, Morisi, Coletti & Catanzaro (1974) have recorded high levels of glutamine in goldfish brain (but not in the liver or kidney) following exposure to ammonia solutions, but no reduction in glutamic acid levels and it was inferred that glutamic acid production was increased. Significantly, brain glutamine levels increased linearly with ambient ammonia concentration up to 10.5 mg l^{-1} as N. Furthermore, on return to clean water the brain glutamine content returned to normal over a period of 2 days, which is in line with the evidence given above for the short-term retention of acclimation.

Effects of phenol and mixtures of ammonia and phenol

Although of diminishing importance, phenol is still commonly found in low concentrations in polluted rivers; a critical review of the literature on the effect of phenol on fish has been made by EIFAC (1972). The toxic action of phenol has been variously attributed to effects on the nervous system, causing paralysis and convulsions, and interference with the respiratory system resulting in asphyxia. Sublethal concentrations have been shown to give rise to a series of pathological effects including necrosis of the gills and increased gill mucus production (Skrapek, 1963; Reichenbach-Klinke, 1965), histopathological changes in heart, liver, spleen and skin tissues (Mitrovic, Brown, Shurben & Berryman, 1968) and destruction of erythrocytes (Halsband & Halsband, 1963; Waluga, 1966).

Studies by Herbert (1962) on the effect of gas liquors on fish showed that phenol and ammonia were additive in their action even though the toxic mechanisms were probably different. Preliminary experiments have been done to investigate further the basis of such joint action (Swift, unpublished data) and to study whether the increased water uptake caused by high ambient ammonia concentrations is accompanied by an increased rate of phenol uptake.

Initial experiments showed that there is no significant change in the UFR of rainbow trout exposed to up to 6 mg l^{-1} phenol; at higher levels inconsistent results were obtained, mainly because the fish became more active as the phenol concentrations neared the lethal level of 9 mg l^{-1}. However, haematocrit values of rainbow trout exposed to 1, 2, 6, 7 and 9 mg l^{-1} phenol for 24 hours were not significantly different from those of the controls, so it is unlikely that the fish were suffering from respiratory stress. Although changes in haematocrit have been reported for bream exposed to phenol (Waluga 1966) more recent work by Kristofferson, Broberg & Oikari (1973) with pike kept in brackish water and exposed to phenol for up to 5 days, showed no such change.

Analysis of muscle from rainbow trout exposed to phenol for 24 hours showed that at concentrations up to 10 mg l^{-1} the tissue levels of phenol were similar to those in the ambient solution, although in lethal solutions the concentrations in the muscle were considerably higher. Exposure for 5 days to 5 mg l^{-1} resulted in a muscle phenol concentration of 6.8 mg kg^{-1}. Further experiments showed that the equilibrium concentration was normally achieved in 2–3 hours (Fig. 8). Also, after the removal of rainbow trout from phenol solutions into clean water the tissue phenol levels returned to normal within a similarly short period. Blood and brain tissue showed similar trends.

Experiments with ammonia–phenol mixtures indicate that the equilibrium

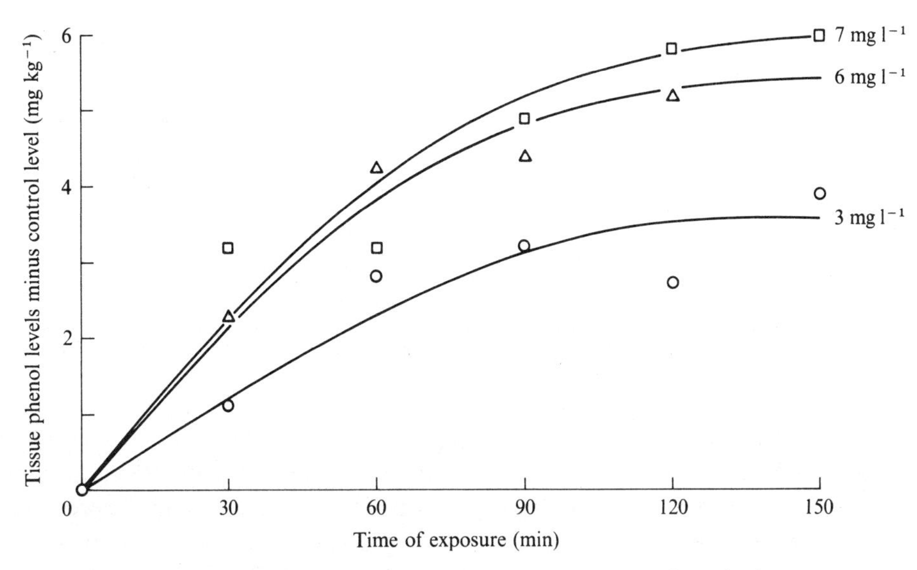

Fig. 8. Increase in rainbow trout muscle phenol concentrations during exposure to phenol solutions. Temperature, 15 °C.

Table 3. *Concentration of phenol in rainbow trout muscle after exposure to phenol or ammonia–phenol mixtures for up to* 24 *hours*

Concentration in solution			
Phenol (mg l^{-1})	Ammonia (mg N l^{-1})	Concentration in muscle ± S.D. (mg kg^{-1})	No. of fish
0	0	1.3 (±0.5)	20
0	5	1.3 (±0.3)	10
0.5	8	2.0 (±0.9)	10
1	8	2.0 (±0.3)	10
3*	12	3.0 (±0.7)	5
5	0	5.7 (±1.5)	9
5	5	6.2 (±1.9)	5
5*	10	9.0 (±0.7)	10
10*	0	13.0 (±0.4)	4
10*	8	13.9 (±0.8)	5

Experimental temperature, 15 °C; pH, 8.0.
* Lethal solutions; fish sampled at 24 hours or at death.

concentration of phenol in muscle is not influenced by elevated ambient ammonia levels even when lethal concentrations of the mixture are reached (Table 3), except in the case of 5 mg phenol l^{-1} and 10 mg NH_3 l^{-1} as N. Furthermore the rate of uptake of phenol into the muscles during the

Table 4. *Mean urine flow rates of rainbow trout exposed to ammonia or ammonia–phenol solutions for 24 hours*

Concentration in solution		Mean urine flow rate (ml kg^{-1} h^{-1})			No. of fish
Phenol (mg l^{-1})	Ammonia (mg N l^{-1})	Pre-test	Test	Post-test	
0	3.5	2.8	5.0	2.6	9
3	3.5	2.8	4.9	2.5	9

Experimental temperature, 17 °C; pH, 8.1.

initial 3-hour exposure period is not influenced by the presence of ammonia. The UFRs of rainbow trout exposed to ammonia solutions are not increased if phenol is also present (Table 4). There is no indication that the presence of ammonia increases the rate of phenol excretion in the urine, although there was a considerable variation in the results obtained. No basis could be found, therefore, for a hypothesis to account for the additive toxic actions of ammonia and phenol; in mixed solutions the effects measured were those pertaining to concentrations of either ammonia or phenol alone and the presence of one did not enhance the effect of the other. The equation used for summing the individual toxicities of each of these poisons in a mixture whereby the concentrations in the solution are expressed as fractions of the lethal threshold concentration and then summed, remains empirical.

Discussion

The experiments which have been described in this paper were carried out to obtain information on some of the basic physiological mechanisms involved in the response by fish to unfavourable environmental conditions. Inevitably, more problems have been raised than have been solved, but some new insights have been obtained, especially in relation to the adaptability of fish to environmental changes and some of the mechanisms which may be involved.

The normal UFR of freshwater fish is a function of body surface permeability and possibly water ingestion by drinking; the literature has been reviewed by Hickman & Trump (1969). An increase in temperature increases the UFR and Lloyd & Orr (1969) showed that for rainbow trout the increase was linear with temperature, the rate at 20 °C being 2.4 times that at 10 °C; this relation is close to the value of 2.2 found for white sucker fish by Mackay & Beatty (1968). It was also noted that very active fish tended to have a higher UFR than those which were less active although, since no quantitative measurements of activity were made, no correlation between these two parameters was possible. However, vigorous activity within a confined apparatus may well have produced integumental damage, particularly to the fins, or affected the mucus layer, leading to a greater uptake of water.

It is well-known that handling of fish prior to experimentation causes diuresis (Hickman & Trump, 1969) and at least a day has to elapse before urine flow and composition return to a steady level. With some active rainbow trout, however, it was noted that a high initial UFR was subsequently followed by a period of very low UFR when the fish was quiescent; for example, in one experiment the initial UFR was 4.9 ml kg^{-1} h^{-1} and after the level of free carbon dioxide had been raised after 2 days the UFR fell from a stable rate of about 3.4 to 0.4 ml kg^{-1} h^{-1} for a short period. The urine produced contained elevated levels of ions, so the reduced UFR was the result of resorption of water by the kidney rather than reduced filtration rate. It is possible that, in this instance, there had been an initial haemoconcentration with subsequent return to normal. Soivio, Westman & Nyholm (1972) have reported an initial elevated haematocrit in rainbow trout after dorsal aorta catheterisation, the value returning to normal within 24 hours. Handling and manipulation may cause an increase in the oxygen consumption of fish and we have shown that respiratory stress caused by hypoxia leads to an increased UFR and possibly haemoconcentration. However, it is also possible that handling leads to increased salt loss through the integument, and the evidence of the acclimation of rainbow trout from high ambient free carbon dioxide levels suggests that an increased UFR can compensate for salt loss, although obviously this can only be a short-term measure.

Laboratory diuresis may be caused, therefore, by any one, or a combination, of these factors, the relative importance of which may vary between fish and between experimental procedures. Increased UFR under these conditions may not be a simple function of increased integumental permeability as the review by Hickman & Trump (1969) would suggest. However, it may be difficult to distinguish between prime and secondary responses. For example, in the preliminary experiments carried out with a modified divided chamber to measure the change in net flux of sodium and chloride ions in the head region of rainbow trout exposed to a hypoxic environment (Lloyd, unpublished data) it was found that at the onset of hypoxia both sodium and chloride net uptake rates were reduced. It is not known whether the onset of hypoxia reduces the active uptake of sodium and chloride by the gills and that this is compensated by a reduction in blood volume with concomitant increased haematocrit, or whether the prime response is the increase in haematocrit with a secondary reduction in net sodium and chloride uptake to avoid an increase in blood osmolarity. These hypotheses will remain tentative until direct measurements are made of the actual change in blood volume under these conditions, taking care that the experimental technique used does not cause such a change to take place.

A period of hypoxia of more than 5 hours' duration causes an increase in body permeability; it is not known why this occurs, particularly at a time when the ventilation rate begins to return to normal. Nevertheless it must lead to a re-examination of the mechanism by which poisons are more harmful to fish in water of low dissolved oxygen content, since increased permeability to water may be accompanied by an increased uptake of toxic chemicals. However, increased permeability to water caused by elevated ambient ammonia concentrations did not increase the uptake of phenol by fish, so the uptake rates may not necessarily be linked. A hypothesis has been put forward which associates the increased toxicity of poisons with increased ventilation rate during periods of hypoxia (Lloyd, 1961), but there has been no experimental confirmation of its validity, although it has become widely accepted. It would seem that further research is required in this field.

No explanation can be given for the diuretic response of fish to high ammonia concentrations; it is not known whether this is associated with the greater integumental permeability or to an increase in water swallowing, although Denis (1966) found that ammonia increased the permeability of tissues in the dog. Because of the high UFRs associated with lethal ammonia levels it is tempting to associate death with renal overloading, but this would not explain mortality in mixtures of ammonia and phenol where the UFRs were lower than in lethal ammonia solutions. Furthermore, post-handling diuresis occasionally reached levels which were associated with mortality in ammonia solutions, but without lethal effect.

An understanding of the physiological response mechanism of fish to all forms of adverse environment factors is essential if the effect of those additional stresses caused by pollutants is to be assessed or perhaps forecast. It is clear that the responses measured so far are inadequate for an explanation of the joint action of pollutants and that other physiological responses have to be investigated. It is important to remember that such investigations will have a bearing not only on the assessment of pollutional stress, but will be of value also in other branches of fisheries management, including intensive fish farming where ammonia, hypoxia and handling are common stressors.

References

Alabaster, J. S., Herbert, D. W. H. & Hemens, J. (1957). The survival of rainbow trout (*Salmo gairdneri*) and perch (*Perca fluviatilis*) at various concentrations of dissolved oxygen and carbon dioxide. *Annals of Applied Biology*, **45**, 177–88.

Basu, S. P. (1959). Active respiration of fish in relation to ambient concentrations of oxygen and carbon dioxide. *Journal of the Fisheries Research Board of Canada*, **16**, 175–212.

Brett, J. R. (1958). Implications and assessments of environmental stress. In *The investigation of fish-power problems* (ed. P. A. Larkin), pp. 69–83. University of British Columbia Press.

Burrows, R. E. (1964). *Effects of accumulated excretory products on hatchery reared salmonids.* United States Bureau of Sport Fisheries and Wildlife, Research Report 66.

Dahlberg, M. L., Shumway, D. L. & Doudoroff, P. (1968). Influence of dissolved oxygen and carbon dioxide on the swimming performance of largemouth bass and coho salmon. *Journal of the Fisheries Research Board of Canada*, **25**, 49–70.

Davis, G. E., Forster, J., Warren, C. E. & Doudoroff, P. (1963). The influence of oxygen concentration on the swimming performance of juvenile Pacific salmon at various temperatures. *Transactions of the American Fisheries Society*, **92**, 111–24.

Denis, G. (1966). Rôle de l'ammoniaque sur la perméabilité des tissus. *Union Médicale du Canada*, **95**, 1453.

Department of the Environment (1973). *Water pollution research, 1972.* HMSO, London.

Doudoroff, P. and Shumway, D. L. (1970). *Dissolved oxygen requirements of freshwater fishes. FAO Fisheries Technical Paper No. 86.* Food and Agricultural Organization of the United Nations, Rome.

Eddy, F. B. & Morgan, R. I. G. (1969). Some effects of carbon dioxide on the blood of rainbow trout *Salmo gairdneri* Richardson. *Journal of Fish Biology*, **1**, 361–72.

EIFAC (European Inland Fisheries Advisory Commission) (1970). *Water quality criteria for European freshwater fish. Report on ammonia and inland fisheries. EIFAC Technical Paper No. 11.* Food and Agricultural Organization of the United Nations, Rome.

EIFAC (European Inland Fisheries Advisory Commission) (1972). *Water quality criteria for European freshwater fish. Report on monohydric phenols and inland fisheries. EIFAC Technical Paper No. 15.* Food and Agricultural Organization of the United Nations, Rome.

EIFAC (European Inland Fisheries Advisory Commission) (1973). *Water quality criteria for European freshwater fish. Report on dissolved oxygen and inland fisheries. EIFAC Technical Paper No. 19.* Food and Agricultural Organization of the United Nations, Rome.

Flis, J. (1968). Anatomico-histopathological changes induced in carp (*Cyprinus carpio* L.) by ammonia water. Part 1: effect of toxic concentrations. Part 2: effects of sub-toxic concentrations. *Acta Hydrobiologica*, **10**, 205–38.

Fry, F. E. J. (1957). The aquatic respiration of fish. In *The physiology of fishes* (ed. M. E. Brown), pp. 1–63. Academic Press, New York & London.

Garey, W. F. & Rahn, H. (1970). Gas tensions in tissues of trout and carp exposed to diurnal changes in oxygen tension of the water. *Journal of Experimental Biology*, **52**, 575–82.

Hall, F. G. (1928). Blood concentration in marine fishes. *Journal of Biological Chemistry*, **76**, 623–31.

Halsband, E. & Halsband, I. (1963). Veränderungen des Blutbildes von Fischen infolge toxischer Schäden. *Archiv für Fischerei Wissenschaft*, **14**, 68–85.

Herbert, D. W. M. (1962). The toxicity to rainbow trout of spent still liquors from the distillation of coal. *Annals of Applied Biology*, **50**, 755–77.

Hickman, C. P. Jr & Trump, B. F. (1969). The kidney. In *Fish physiology* (ed. W. S. Hoar & D. J. Randall), vol. 1, pp. 91–239. Academic Press, New York & London.

Hughes, G. M. (1973). Respiratory responses to hypoxia in fish. *American Zoologist*, **13**, 475–89.

Hunn, J. B. (1969). Chemical composition of rainbow trout urine following acute hypoxic stress. *Transactions of the American Fisheries Society*, **98**, 20–2.

Hunn, J. B. & Willford, W. A. (1970). The effects of anesthetization and urinary bladder catheterization on renal function of rainbow trout. *Comparative Biochemistry and Physiology*, **33**, 805–12.

Kristofferson, R., Broberg, S. & Oikari, A. (1973). Physiological effects of a sublethal concentration of phenol in the pike (*Esox lucius* L). in pure brackish water. *Annales Zoologici Fennici*, **10**, 392–7.

Levi, G., Morisi, G., Coletti, A. & Catanzaro, R. (1974). Free amino acids in fish brain: normal levels and changes upon exposure to high ammonia concentrations *in vivo* and upon incubation of brain slices. *Comparative Biochemistry and Physiology*, **49A**, 623–36.

Lloyd, R. (1960). The toxicity of zinc sulphate to rainbow trout. *Annals of Applied Biology*, **48**, 84–94.

Lloyd, R. (1961). Effect of dissolved-oxygen concentrations on the toxicity of several poisons to rainbow trout (*Salmo gairdneri* Richardson). *Journal of Experimental Biology*, **38**, 447–55.

Lloyd, R. & Jordan, D. H. M. (1964). Some factors affecting the resistance of rainbow trout (*Salmo gairdneri* Richardson) to acid waters. *International Journal of Air and Water Pollution*, **8**, 393–403.

Lloyd, R. & Orr, L. D. (1969). The diuretic response by rainbow trout to sub-lethal concentrations of ammonia. *Water Research*, **3**, 335–44.

Lloyd, R. & White, W. R. (1967). Effect of high concentrations of carbon dioxide on the ionic composition of rainbow trout blood. *Nature, London*, **216**, 1341–2.

Mackay, W. C. & Beatty, D. D. (1968). The effect of temperature on renal function in the white sucker fish *Catostomus commensonii*. *Comparative Biochemistry and Physiology*, **26**, 235–45.

McLeay, D. J. & Brown, D. A. (1974). Growth stimulation and biochemical changes in juvenile coho salmon (*Oncorhynchus kisutch*) exposed to bleached kraft pulpmill effluent for 200 days. *Journal of the Fisheries Research Board of Canada*, **31**, 1043–9.

Maetz, J. & Garcia Romeu, F. (1964). The mechanisms of sodium and chloride uptake by the gills of a freshwater fish *Carassius auratus*. III. Evidence for NH_4^+/Na^+ and HCO_3^-/Cl^- exchanges. *Journal of General Physiology*, **47**, 1209–27.

Ministry of Technology (1967). *Water pollution research, 1966*. HMSO, London.

Mitrovic, V. V., Brown, V. M., Shurben, D. G. & Berryman, M. M. (1968). Some pathological effects of sub-acute and acute poisoning of rainbow trout by phenol in hard water. *Water Research*, **2**, 249–54.

Mount, D. I. & Stephan, C. E. (1967). A method for establishing acceptable toxicant limits for fish: malathion and butoxyethanol ester of 2,4-D. *Transactions of the American Fisheries Society*, **96**, 185–93.

Ostroumova, I. N. (1954). Conditions of trout blood at adaptation to different oxygen and salt conditions of water. *Izvestiya Vsesoyuznogo (gos). nauchno-issled. Institut Ozernogo i Rechnogo Rybnogo Khozyaistva*, **58**, 27–36.

Post, G., Shanks, W. E. & Smith, R. R. (1965). A method of collecting metabolic excretions from fish. *Progressive Fish Culturist*, **27**, 108–11.

Privolnev, T. I. (1954). *Physiological adaptations of fishes to new conditions of existence.* Fisheries Research Board of Canada, Translation Series, No. 422.

Randall, D. J. (1970). Gas exchange in fish. In *Fish physiology* (ed. W. S. Hoar & D. J. Randall), vol. 4, pp. 253–92. Academic Press, New York & London.

Reichenbach-Klinke, H.-H. (1965). Der Phenolgehalt des Wassers in seiner Auswirkung auf den Fischorganismus. *Archiv für Fischerei Wissenschaft*, **16**, 1–16.

Saunders, R. L. (1962). The irrigation of the gills of fishes. II. Efficiency of oxygen uptake in relation to respiratory flow activity and concentrations of oxygen and carbon dioxide. *Canadian Journal of Zoology*, **40**, 817–62.

Shepard, M. P. (1955). Resistance and tolerance of young speckled trout (*Salvelinus fontinalis*) to oxygen lack, with special reference to low oxygen acclimation. *Journal of the Fisheries Research Board of Canada*, **12**, 387–446.

Skrapek, K. (1963). Toxicity of phenols and their detection in fish. Abstract in: *Public Health Engineering Abstract* (1964), **44**, 272.

Smith, L. S. (1966). Blood volumes of three salmonids. *Journal of the Fisheries Research Board of Canada*, **23**, 1439–46.

Soivio, A., Westman, K. & Nyholm, K. (1972). Improved method of dorsal aorta catheterization: haematological effects followed for three weeks in rainbow trout (*Salmo gairdneri*). *Finnish Fisheries Research*, **1**, 11–21.

Sprague, J. B. (1971). Measurement of pollutant toxicity to fish. III. Sublethal effects and 'safe' concentrations. *Water Research*, **5**, 245–66.

Stevens, E. D. (1968). The effect of exercise on the distribution of blood to various organs in rainbow trout. *Comparative Biochemistry and Physiology*, **25**, 615–25.

Swift, D. J. & Lloyd, R. (1974). Changes in urine flow rate and haematocrit value of rainbow trout (*Salmo gairdneri* Richardson) exposed to hypoxia. *Journal of Fish Biology*, **6**, 379–87.

Van Dam, L. (1938). On the utilization and regulation of breathing in some aquatic animals. PhD dissertation, Gröningen University.

Waluga, D. (1966). Phenol-induced changes in the peripheral blood of the bream (*Abramis brama L*). *Acta Hydrobiologica*, **8**, 87–95.

Wood, S. C. & Johansen, K. (1972). Adaptation to hypoxia by increased HbO_2 affinity and decreased red cell ATP concentration. *Nature New Biology*, **237**, 278–9.

E.D.S.CORNER & R.P.HARRIS (Part I)
K.J.WHITTLE & P.R.MACKIE (Part II)

Hydrocarbons in marine zooplankton and fish

In recent years detailed analyses of seawater, as well as marine species ranging from micro-organisms to vertebrates, have demonstrated the presence of hydrocarbons, both aliphatic and aromatic, which are similar in nature to components of petroleum and petroleum products. Also present, however, are other hydrocarbons that are characteristic of living organisms.

In the open sea concentrations of hydrocarbons are generally low and the origin of these compounds is not always easily determined. By contrast, in areas affected by massive contamination the hydrocarbons may be present in high concentrations and can be directly related to the particular source of pollution. These two different situations have prompted studies of both short-term and long-term effects of petroleum hydrocarbons, the former being principally concerned with toxicity, tainting and behavioural responses, and the latter with uptake, metabolism and release or possible accumulation.

This two-part paper reviews the background to these studies and then discusses pertinent experiments that are still in progress, the first section being mainly concerned with zooplankton and the second with fish.

PART I. ZOOPLANKTON

Hydrocarbon levels

In sea water

Estimates of 'dissolved' and particulate hydrocarbons in the sea have been made by several workers (see Table 1), 'dissolved' material being generally defined as that passing through a particular filter, usually a Millipore membrane of 0.45 μm pore size. The filtrate is extracted with a solvent, such as chloroform or methylene chloride, and the extract examined either for total hydrocarbon content, using fluorescence spectroscopy, or for a particular group of compounds after chromatographic separation. Gordon, Keizer & Dale (1974) have drawn attention to several possible sources of error. Thus, the levels of hydrocarbons in the samples often represent the lowest limits

Table 1. *Levels of hydrocarbons in seawater*

Region	Depth	Filter	Concentration (ppb) Particulate	Dissolved	Reference
Outer reaches of Chedabucto Bay (Nova Scotia)	1 m	0.45 μm	5–16[a]	15–90[a]	Levy (1971)
Gotland Deep (Baltic)	20–200 m	Whatman GF/C	0.5–2.3[b]	48–64[b]	Zsolnay (1971)
Baffin Bay (Texas)	—	0.3 μm	70[c]	180[c]	Jeffrey (1970)
Cap Ferrat (Mediterranean)	50 m	0.45 μm	—	20.7–81.9	Copin & Barbier (1971)
Brest	Surface	0.45 μm	—	137	Barbier *et al.* (1973)
Villefranche	50 m	0.45 μm	—	75	Barbier *et al.* (1973)
Roscoff	Surface	0.45 μm	—	46	Barbier *et al.* (1973)
Open sea off W. Africa	50–4500 m	0.45 μm	—	10–95	Barbier *et al.* (1973)
N.W. Atlantic (Halifax–	0–3 mm	—		20.4[d]	Gordon *et al.* (1974)
Bermuda	1 m			0.8[d]	Gordon *et al.* (1974)
section)	5 m			0.4[d]	Gordon *et al.* (1974)
	5 m			Nil	Gordon *et al.* (1974)

[a] As Bunker C oil equivalents; [b] as carbon; [c] unsaturated hydrocarbons; [d] as Venezuelan oil equivalents.

of the analytical methods; contamination of the samples can occur during collection; and naturally occurring hydrocarbons cause errors if fluorescence spectroscopy is used with crude oil as a standard. The data in Table 1 refer to different sea areas and deal with different classes of compound: nevertheless, one general finding is that concentrations of 'dissolved' hydrocarbons are much higher than the levels in particulate form. There is also evidence from the work of Gordon *et al.* (1974) that in seawater collected from the Halifax–Bermuda section (N.W. Atlantic) hydrocarbons are concentrated at the surface. Similar findings have been made by Whittle, Mackie, Hardy & McIntyre (1973) for Scottish waters and have been confirmed for other sea areas, both open-sea and coastal, examined subsequently (Hardy, Whittle & Mackie, unpublished data).

A comprehensive analysis of dissolved hydrocarbons is that of Barbier, Joly, Saliot & Tourres (1973) who examined chloroform extracts of filtered 100-l samples of coastal and oceanic waters by first separating the unsaponifiable material and then analysing it by means of thin-layer and gas–liquid chromatography, followed by mass spectrometry. Some of their findings are included in Table 1: in addition, detailed examination of the 'Brest' sample showed that it contained normal and branched paraffins (51 %), cycloalkanes

(25 %), mono-cyclic (18 %), bi-cyclic (3.5 %) and poly-cyclic aromatics (2.5 %). Thus, the concentrations of dissolved bi-cyclic compounds can be estimated as 4.8 ppb ($\mu g\ l^{-1}$).

In plankton

The hydrocarbons in plankton have been studied in some detail by Blumer and co-workers (Blumer, Robertson, Gordon & Sass, 1969; Blumer, Mullin & Thomas, 1970; Blumer, Guillard & Chase, 1971). Relatively high concentrations of various isoprenoid hydrocarbons, particularly pristane, were found in calanoid copepods (1–3 % total lipid) and the polyunsaturated hydrocarbon, heneicosahexaene, in *Rhincalanus nasutus* (0.1–0.5 % total lipid). Pristane is thought to be derived from phytol, a breakdown product of chlorophylls in plant diets; heneicosahexaene could be obtained from plant diets directly, as it has been found in several species of marine phytoplankton. Straight-chain saturated paraffins (e.g. n-heptadecane) are also present in trace amounts; and recently Whittle, Mackie & Hardy (1974*a*) found that C_{18}–C_{34} n-alkanes accounted for 3.3 % fresh weight of plankton samples collected from the Clyde and the North Sea. By contrast, cycloalkanes and aromatic compounds, characteristic of both crude oil and refined petroleum products, are not normally present and therefore the implication of oil-pollution as a source of hydrocarbons in marine plankton is usually based on evidence of high levels of aromatic compounds in the test samples (see, for example, Morris, 1974). It should be noted, however, that poly-cyclic aromatic hydrocarbons of the benzo[*a*]pyrene type have been identified by ultraviolet fluorescence in trace amounts in plankton samples (Mallet & Lami, 1964) collected from inshore areas near the French Channel coast, values being in the range 100–350 $\mu g\ kg^{-1}$ dry weight. The presence of these hydrocarbons does not necessarily indicate oil pollution: land drainage and human sewage are other possible sources. It is also worth noting that derivatives of aromatic hydrocarbons can occur naturally in some marine invertebrates, well-known examples being quinone compounds based on naphthalene, anthracene and benzo[*a*]pyrene (see review by Scheur, 1973). Again, although aromatic hydrocarbons themselves may occur in marine animals as the result of pollution, they could also arise, particularly in animals that dwell in muds and sediments, from synthesis by anaerobic bacteria (Zobell, 1971).

Toxicity

The toxicity of crude oil or refined petroleum products to zooplankton has so far received little attention apart from the work of Mironov (1969) and Barnett & Kontogiannis (1975). Mironov examined the survival rates of several species of copepod from the Black Sea that fed in seawater containing different amounts of various oils and concluded that effects on survival were manifest at the 1.0 ppm (mg l^{-1}) level. This observation has been quoted (Mileikovsky, 1970; Dunning & Major, 1974) as showing that oil is very toxic. However, the difference in survival times of the control animals and those exposed to the lowest oil concentrations in Mironov's experiments was marginal; moreover, the populations used seem to have been under stress in that 50 % of the controls died in less than a week. Barnett & Kontogiannis (1975) measured the toxicities of various crude oil fractions using the harpacticoid copepod *Tigriopus californicus* and found, as might be expected, that this intertidal animal was much more resistant than the species tested by Mironov (1969). Thus, the concentration of diesel oil had to exceed 87 ppm and that of kerosene 83 ppm before the treated animals died any faster than the controls.

Other workers have been less concerned with the effects of oil on zooplankton than with the effects of zooplankton on oil. For instance, Freegarde, Hatchard & Parker (1971) found that the copepod *Calanus finmarchicus*, when transferred to seawater containing the diatom *Phaeodactylum tricornutum* and a fine suspension of Kuwait crude oil at a concentration of 2–10 ppm, ingested the oil and released it in faecal pellets. A similar observation was made by Conover (1971) using *C. finmarchicus* and *Temora longicornis* that had been feeding in the sea on a weathered sample of Bunker C oil at a concentration of 0.01–0.02 ppm: on average, the oil accounted for 2 % of the dry weight of the faeces, with a maximum of 7 %. Freegarde *et al.* (1971) and Conover (1971) concluded that zooplankton, by incorporating oil into faecal pellets slightly heavier than seawater and containing bacteria that might hasten decomposition of hydrocarbons, could be effective in immobilising a substantial fraction of an oil spill. The extent to which the animals can do this, however, obviously depends on the toxicity of the oil; and it is worth noting that the concentrations of crude oil used in the short-term experiments (18 h) of Freegarde *et al.* (1971) were within a range that Mironov (1969) claimed was toxic to zooplankton when tested over longer periods (3–7 days). In a recent study (Spooner & Corkett, 1974) oil at a concentration of 10 ppm was found to cause a substantial reduction in faecal pellet production by *Calanus* and other species of zooplankton over a period of 24 h.

An important factor influencing the toxicity of an oil is the size and nature

Table 2. *Toxicities and aromatic hydrocarbon contents of different oils*

	S. Louisiana crude	Kuwait crude	No. 2 Fuel oil	Bunker C oil
Concn (ppm) of oil causing 50 % death of *Palaemonetes pugio* in 24 h	>16.8	>10.2	4.4	3.2
Total dissolved hydrocarbons (ppm)	23.76	21.65	5.28	1.36
Dissolved bi- and tri-cyclic aromatics (ppm)	0.31	0.08	2.00	0.94

Data from Anderson *et al.* (1974).

of the water-soluble fraction. This includes the light aromatic compounds, such as benzene and toluene, which are rapidly lost by weathering (see, for example, Frankenfeld, 1973); but also present are naphthalene and its alkyl derivatives (1- and 2-methylnaphthalene, dimethyl- and trimethyl-naphthalene), which are more persistent. Boylan & Tripp (1971) noted that kerosene was more toxic than crude oil to marine fish and ascribed this difference to the relative quantities of naphthalene in the aqueous extracts (32.3 ppb from crude oil and 159.9 ppb from kerosene). Anderson *et al.* (1974) reached a similar conclusion after testing four kinds of oil for toxicity to a number of marine crustaceans and fish. Typical of their data are those shown in Table 2, from which it is clear that the high toxicities of No. 2 Fuel Oil and Bunker C to the prawn *Palaemonetes pugio* are related to the much higher proportions of bi- and tri-cyclic aromatic compounds in the water-soluble fractions of these oils.

It is interesting to examine the toxicity data of Mironov (1969), bearing in mind the importance of water-soluble compounds in determining the toxicity of an oil. In his experiments with zooplankton Mironov showed that a definite toxic effect occurred with fuel oil used at a concentration of 10 ppm. Working with a freshly prepared oil-in-water dispersion of No. 2 Fuel Oil, Anderson *et al.* (1974, table 4) found that roughly 1 % of the hydrocarbons were in the aqueous phase, of which nearly half consisted of bi- and tri-cyclic aromatic compounds. Oils differ considerably in hydrocarbon content, but assuming that the data of Anderson *et al.* also apply to the fuel oil used by Mironov, the toxic effects he detected using 10 ppm fuel oil could have been caused by bi- and tri-cyclic compounds at a concentration totalling 50 ppb. This value is an order of magnitude higher than the levels of these compounds detected in the sea off Brest by Barbier *et al.* (1973).

A further point concerns the toxicity of weathered oil: it is now well-established that during the weathering of crude oil, oxidation processes can lead to the formation of thiacyclanes (sulphoxides) and alkylphenols (see, for example, Burwood & Spears, 1974). These substances may affect the toxicity of the oil to zooplankton, but so far there seems to have been no work done on this aspect of the problem.

Uptake and release

Studies on the uptake, retention and release of hydrocarbons have been made with several species of marine invertebrate. Generally, it has been found that uptake from solution in seawater is rapid, but after the animals are transferred to uncontaminated seawater large fractions of the hydrocarbon are quickly lost (Lee, Sauerheber & Benson, 1972; Anderson *et al.*, 1974). However, a small fraction remains that decreases very slowly. For example, Stegeman & Teal (1973) found that specimens of the American oyster, *Crassostrea virginica*, exposed to a concentration of 106 μg l^{-1} of No. 2 Fuel Oil for 7 weeks accumulated 334 μg per g fresh tissue: following transfer of the animals to fresh seawater, 90 % of the hydrocarbon content was lost linearly over a further 2 weeks, after which the rate of loss slowed dramatically, levelling off at about 10 % of the original concentration of hydrocarbon. A further finding was that the quantity of hydrocarbons accumulated by oysters during an exposure period of 7 weeks was greater in animals that contained a higher proportion of body lipid. Conceivably, the small quantities of fossil-fuel hydrocarbons retained by the animals over long periods are concentrated in some more stable lipid fraction from which they can be released only very slowly. On the other hand, it is also necessary to consider whether some hydrocarbons, particularly aromatic compounds, undergo biochemical changes in the animals and are retained as metabolites (Lee, 1975).

Metabolism

A detailed account of the metabolism of poly-cyclic aromatic hydrocarbons in mammals has appeared elsewhere (Corner, 1975): a summary of the main results of in-vitro studies with naphthalene is given in Fig. 1.

The formation of hydroxy derivatives is characteristic of the metabolism of numerous aromatic hydrocarbons and, generally speaking, the more complex the hydrocarbon is the greater the number of derivatives it forms: for example

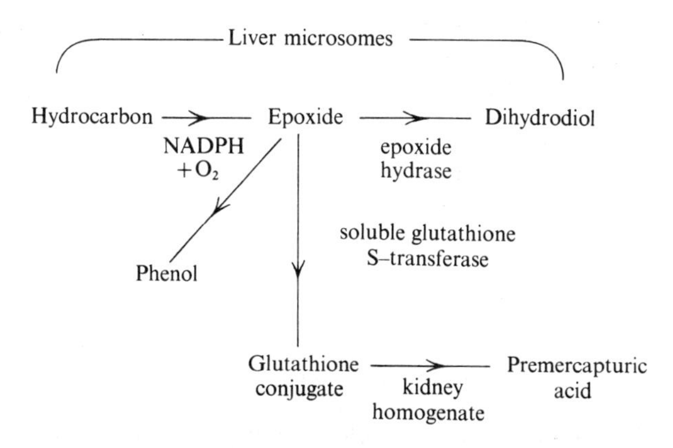

Fig. 1. Summary of in-vitro studies of naphthalene metabolism.

Holder *et al.* (1974) list sixteen reported metabolites of the carcinogen benzo[*a*]pyrene. Epoxides are formed as intermediates in the metabolism of many aromatic hydrocarbons, and recent in-vitro studies by Booth & Sims (1974) have demonstrated that epoxide formation can occur with dihydrodiols as substrates, as well as with hydrocarbons.

Studies of hydrocarbon metabolism *in vivo* have shown that the urines of the animals contain phenols and dihydrodiols both in the free state and as conjugates with sulphuric and glucuronic acids. Glutathione conjugates are converted into acetyl-cysteine derivatives, or 'premercapturic acids', before being excreted in the urine, but occur unchanged, together with the hydroxylated derivatives, in the bile.

Investigations of the fate of aromatic hydrocarbons in certain invertebrates have indicated that these animals can convert the compounds into metabolites similar to those produced by mammals. For example, Terrière, Boose & Roubal (1961) showed that the housefly *Musca domestica* converted naphthalene into nine of the compounds identified as metabolites of the hydrocarbon in rats: more recent work (Corner, Kilvington & O'Hara, 1973) has shown that naphthalene is converted in the crab *Maia squinado* into hydroxylated derivatives as well as an acetyl-cysteine conjugate.

So far, in-vitro studies of the metabolism of aromatic compounds in invertebrates seem to have been mainly concerned with attempts to detect the substrate-inducible enzyme known as mixed-function oxygenase. This enzyme, which has a requirement for molecular oxygen, NADPH and the haemoprotein cytochrome P-450 (Lu *et al.*, 1972), is generally involved in the metabolism of 'foreign' organic compounds of many kinds and is therefore relevant to aspects of both ecology and toxicology (Whitlock & Gelboin, 1974). An enzyme involved in the hydroxylation of naphthalene and using cytochrome P-450 as a terminal oxidase has been found in microsomal preparations of the housefly by Morello, Bleecker & Agosin (1971): such findings led Elmamlouk, Gessner & Brownie (1974) to carry out a similar investigation with the lobster *Homarus americanus*. Cytochrome P-450 was detected in a microsomal fraction of lobster hepatopancreas, but further data (unpublished) indicated that the preparation was unable to oxidise substrates such as biphenyl, ethylmorphine or aniline.

Current experiments with zooplankton

Because of the importance of copepods in the marine food web, recent studies have been made of the uptake and release by different species of the bi-cyclic aromatic compound naphthalene. A brief description of some of the preliminary findings is now given.

Experiments using ^{14}C-labelled naphthalene in solution

Fig. 2 shows the level of radioactivity accumulated by *Eurytemora affinis* immersed for 24 h in seawater containing ^{14}C-labelled naphthalene at a concentration of 0.96 ppm; also, the subsequent fall in level of radioactivity when the treated animals were transferred to fresh seawater. Initially the rate of decrease was rapid, over half being lost in 1 day. After this, however, the rate of loss slowed considerably and the small quantity of radioactivity detected after 6 days persisted to the end of the experiment (9 days).

Similar experiments were carried out using *Calanus helgolandicus*. In studies with males, 80 % of the radioactivity accumulated by animals immersed for 3 h in a concentration of 0.78 ppm ^{14}C-labelled naphthalene was lost 5 h after they had been transferred to fresh seawater. The quantity of retained radioactivity levelled off after 6 days at approximately 1 % of the original concentration and this amount was still detectable at the end of 2 weeks.

A recent study by Lee (1975) has shown that the copepod *Calanus plumchrus*, when immersed in seawater containing 1.25 ppb [^{3}H]benzo[*a*]pyrene, continued to accumulate the hydrocarbon over 3 days, the maximum level attained being 2.2 ng per animal: no further uptake occurred when the exposure period was extended to 10 days. Exposed animals transferred to clean seawater lost radioactivity steadily over a period of 2 weeks, but a small amount (0.01 ng per animal) was still present at the end of the experiment (17 days). Similar experiments with the boreal–arctic species *C. hyperboreus* showed that trace amounts of the hydrocarbon were retained after a depuration period of 28 days.

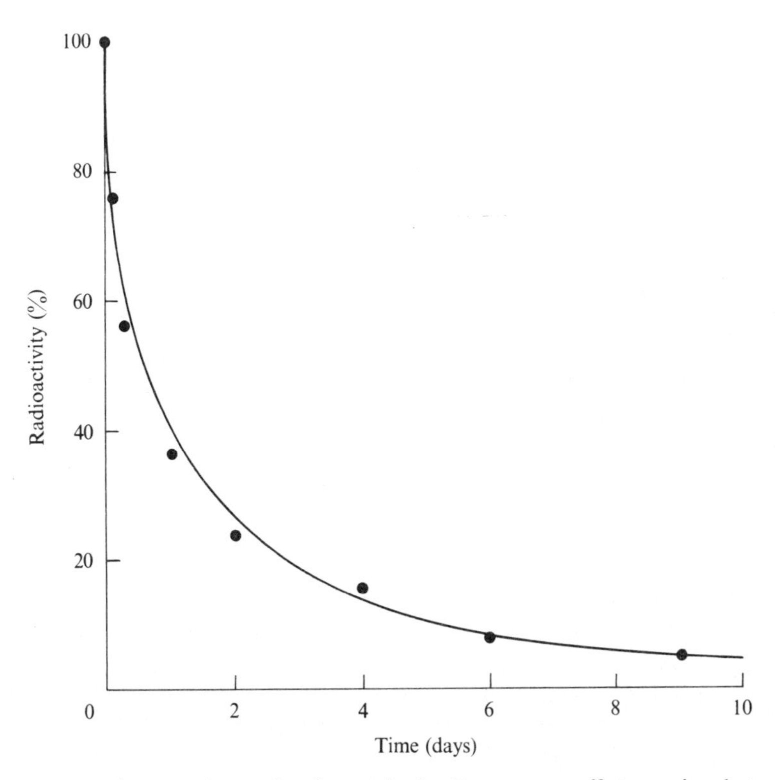

Fig. 2. Release of radioactivity by *Eurytemora affinis* previously treated with ^{14}C-labelled naphthalene.

Experiments using plant cells as a diet

It seemed possible that the rapid loss of naphthalene during the early stages of depuration that followed uptake of the hydrocarbon from solution was simply the result of desorption from sites on the surfaces of the animals. Further experiments were therefore designed in which naphthalene could be administered to the animals by way of the diet (the food in this case being cells of the diatom *Biddulphia sinensis*), earlier studies (Corner, Head & Kilvington, 1972) having shown that *Calanus* feeds actively on this diet, capturing nearly 50 % of the body nitrogen daily at cell concentrations in excess of 10000 l^{-1}.

In the first stage of the experiment, *Biddulphia* cells were immersed in sea-water that had been filtered through a Sartorius membrane (0.2 μm pore size) and contained ^{14}C-labelled naphthalene at a concentration of 300 $\mu g\ l^{-1}$. After 3 days in this solution, the cells were removed and the radioactivity in them was measured.

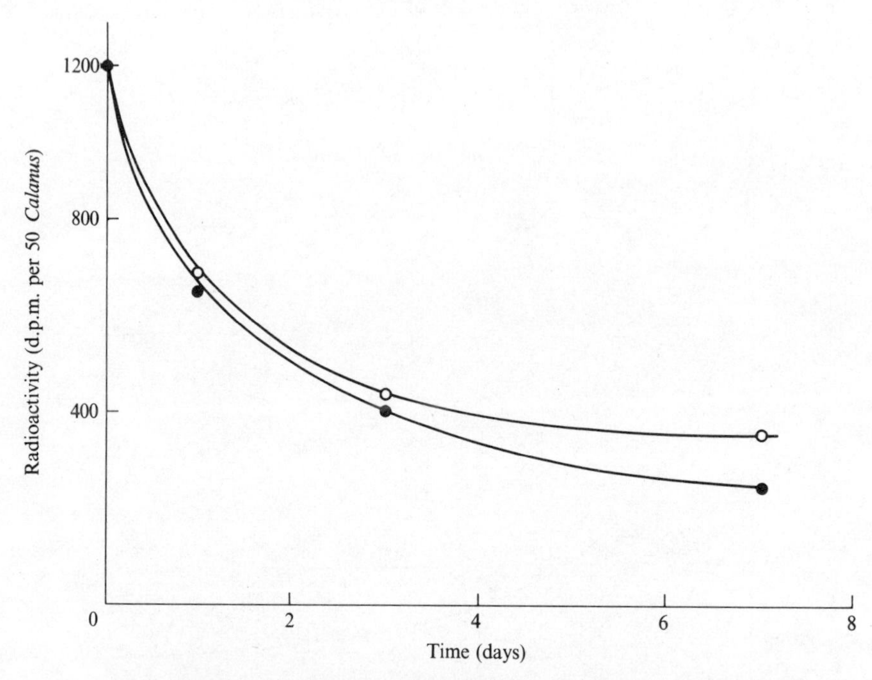

Fig. 3. Release of radioactivity by *Calanus helgolandicus* previously fed on *Biddulphia* treated with ^{14}C-labelled naphthalene. ○, starved during depuration; ●, fed during depuration.

The cells were then fed to adult female *Calanus* in a rotating column device that has been described elsewhere (Corner *et al.*, 1972). The amount of radioactivity captured by each animal was estimated from the number of cells it removed and the average level of radioactivity in these cells during the feeding experiment. Subsequently, the animals were transferred to fresh seawater containing unlabelled *Biddulphia* cells on which they fed for 30 min to clear the guts of radioactive material. Some were then starved and others were given a further diet of unlabelled *Biddulphia*. It is clear from the data in Fig. 3 that both groups of *Calanus* lost radioactivity that had been accumulated from the diet and had presumably entered the tissues of the animals.

Comparison of uptake from solution and the diet

Shown in Fig. 4 is the relationship between levels of radioactivity, as naphthalene equivalents, accumulated by *Calanus* immersed 24 h in seawater containing a wide range (0.01–100 ppb) of concentrations of ^{14}C-labelled naphthalene. Concentrations of radioactivity in dietary form, together with

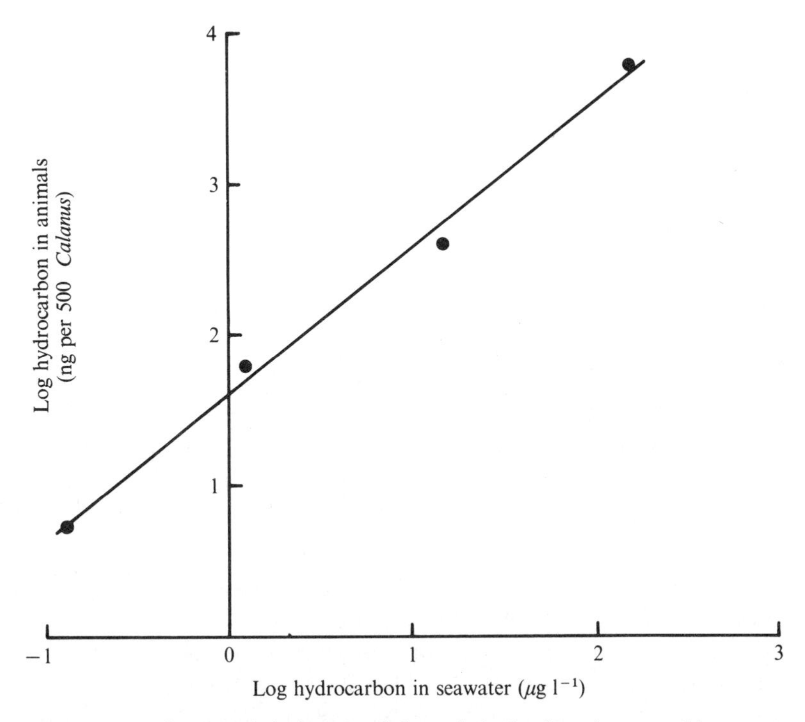

Fig. 4. Levels of naphthalene in *Calanus helgolandicus* immersed in different concentrations of the hydrocarbon in seawater.

Table 3. *Uptake experiments using* Calanus *feeding on* Biddulphia *cells*

Radioactivity accumulated (ng naphthalene per animal)	Hydrocarbon concentration		Concentration ratio dissolved : diet
	As plant cells ($ng\,l^{-1}$)	In seawater ($ng\,l^{-1}$)	
0.25	75.3	3160	42
0.48	86.0	6310	74

Each experiment was carried out with 700 animals. Average concentration of *Biddulphia* cells, 65000 l^{-1}. Hydrocarbon concentration in seawater taken from Fig. 4.

levels in *Calanus* after feeding for 24 h, are shown in Table 3, again expressed as naphthalene equivalents.

The concentrations of naphthalene in solution enabling animals to accumulate directly from seawater the same level of hydrocarbon as those grazing labelled *Biddulphia* cells were calculated from Fig. 4 and are included in Table 3, the last column of which shows the ratio of the concentration of naphthalene in solution to the concentration as plant cells needed to give rise to the same amount of hydrocarbon in the animals.

Evidence for metabolism in zooplankton

Lee (1975) reports that all the crustacean species he examined in his recent studies using zooplankton possessed the ability to metabolise hydrocarbons, the compounds being converted into hydroxylated derivatives. For example, in experiments with *C. plumchrus* animals exposed to benzo[*a*]pyrene at a concentration of 1 ppb for 3 days contained about one-third of the accumulated radioactivity as metabolites.

In the present work with *C. helgolandicus* attempts were made to see whether the radioactivity rapidly lost during the early stages of depuration was released by the animals as naphthalene or its metabolites. The data in Table 4 are the results of six experiments in which the amounts of naphthalene released, measured directly by gas–liquid chromatography (GLC) using n-pentane extracts of seawater containing the animals, are compared with the quantities of the hydrocarbon calculated from the total levels of radioactivity found in this seawater. Average values from the six experiments are in close agreement (31.3 ng estimated from radioactivity; 32.4 ng identified as naphthalene). Thus, there was no evidence that the hydrocarbon had been metabolised. However, these experiments were done under conditions different from those used by Lee (1975): a much shorter exposure period was used (3 h instead of 3 days) together with a much higher concentration of hydrocarbon (0.96 mg naphthalene compared with 1.0 μg benzo[*a*]pyrene). This latter difference seemed particularly important in that the high levels of hydrocarbon used in the present study might have 'swamped' metabolic systems involved in the hydroxylation of the compound. It therefore seemed advisable to carry out further experiments in which animals were exposed to much lower levels of the hydrocarbon. Moreover, it was important to recognise that if metabolism did occur, previous work with *Maia* (Corner *et al.*, 1973) indicated that the products formed were likely to be water-soluble and therefore readily lost from the animals. Accordingly, examination of the tissues of the animal might not be sufficient: it was also necessary to investigate excretion products. Again, in experiments in which the hydrocarbon is taken up directly from solution in seawater it need not necessarily penetrate into the tissues of the animal, but could remain at sites on the surface where attached bacteria might be responsible for any metabolism.

It seemed that these problems could be best overcome by following as

Table 4. *Depuration studies using* Calanus helgolandicus

% Composition of sample			Level of naphthalene in seawater (mg l^{-1})	Naphthalene taken up in 3 h (ng per animal)	% radioactivity released after 4 h in fresh seawater	Naphthalene released (ng per animal)	
♀	♂	Stage V				Calculated from radioactivity in seawater	Estimated by GLC
55	22	23	0.96	31.7	87.0	27.6	25.0
55	22	23	0.94	39.0	94.2	36.0	31.8
70	16	14	1.08	33.9	82.5	27.0	32.0
70	16	14	0.98	42.5	73.8	31.2	34.0
67	30	3	0.96	30.2	89.2	27.0	24.8
67	30	3	0.83	56.6	70.0	38.8	46.8

Each experiment was carried out with 800 animals.

closely as possible the pattern of experiments previously done using mammals: that is, to administer a 'dose' of the hydrocarbon to *Calanus* by way of the diet and then to look for evidence of released metabolites in the seawater during a period of depuration. The test animal, *C. helgolandicus*, feeds actively on phytoplankton but, obviously, living diets could not be used in the present study because of the possibility that the hydrocarbon might be converted into other compounds in the plant cells or by the bacteria associated with them. It was therefore necessary to use a non-living bacteria-free diet, in selecting which use was made of an earlier finding by Corner, Head, Kilvington & Marshall (1974) that *Calanus* will feed on dead nauplii of the barnacle *Elminius modestus*.

Experiments with non-living diets. Dead nauplii were sterilised, either by autoclaving or by treatment with ultraviolet light, and then immersed in a solution of ^{14}C-labelled naphthalene in autoclaved seawater. Having accumulated the hydrocarbon the nauplii were transferred to autoclaved seawater that partially filled a glass column and contained 100 female *Calanus* that had been treated with antibiotics. The column was filled with seawater and then continuously rotated during a 24-h period of feeding in order to keep the food in suspension. The data in Table 5 show the average concentration of hydrocarbon in suspension in the column during the experiment and the level of radioactivity (as pg naphthalene per animal) accumulated during feeding. The *Calanus* were next transferred to autoclaved seawater for a depuration period of 24 h, after which the total content of radioactivity and the amount present as unchanged naphthalene were determined in both the animals and the seawater. The results, shown in Table 5, demonstrate clearly that at the

Table 5. *Metabolism studies using dead* Elminius *nauplii as a diet*

Experiment No.	Naphthalene concentration as *Elminius* nauplii (ng l^{-1})	Naphthalene level after 24-h feeding (pg per *Calanus*)	Naphthalene level after 24-h depuration (pg per *Calanus*)			Amount of radioactivity released as naphthalene (pg per *Calanus*)		
			Calculated	Identified	%	Calculated	Identified	%
1	46	50	34	33	97	16	3.7	23
2	55	93	55	51	93	38	11.5	30
3	63	61	27	27	100	34	9.8	29

Nauplii were sterilised by autoclaving (experiments 1 and 3) or by ultraviolet treatment (experiment 2). Average concentration of nauplii, 10000 l^{-1}. Each experiment carried out with 400 *Calanus*.

end of the depuration period practically all the radioactivity remaining in the animals was still present as naphthalene. By contrast, however, substantial quantities of the radioactivity released into the seawater were no longer in the form of the hydrocarbon, an observation consistent with the view that the compound could have been metabolised.

Conclusions

One purpose of Part I of this paper has been to draw attention to certain gaps in our knowledge concerning the effects of fossil-fuel hydrocarbons on zooplankton. Filling such gaps will require further laboratory experiments, and future work of a useful kind might cover the following aspects.

(1) Studies of the levels of hydrocarbons in the sea have tended to show that more of these compounds are present in the 'dissolved' form than as particulate material. Volatilisation, biodegradation and photo-oxidation are likely to remove considerable quantities of the dissolved compounds and concentrations so far detected, even in coastal areas, seem to be substantially lower than those needed to produce a toxic effect. However, although the concentrations in particulate form are sometimes low compared with those in solution, because of recent evidence that the dietary pathway into the animals may be quantitatively far more important than direct uptake from solution there is a strong case for examining the toxicities of fossil-fuel hydrocarbons, especially bi-cyclic aromatic compounds, administered through the diet.

(2) There is now evidence that certain fossil-fuel hydrocarbons, particularly poly-cyclic aromatic hydrocarbons, can be metabolised by some species of zooplankton and this could be an important factor in the release of such

compounds from the animals. Nevertheless, more experiments, especially of a kind designed to exclude bacterial contamination, will be needed to establish whether a wide variety of zooplanktonic animals retain these hydrocarbons unchanged, or in the form of metabolites.

(3) Although much of the evidence obtained so far indicates that hydrocarbons assimilated by zooplankton, either directly from solution or from their food, are rapidly released once the animals are transferred to fresh seawater, no attempts have been made to correlate uptake and release with lipid content, which could be an important factor influencing both processes and one which could vary considerably from species to species and in any one species with season.

(4) Possible toxic effects of the small amounts of various hydrocarbons retained by zooplankton over long periods deserve detailed study using multiple-generation laboratory cultures that will allow long-term effects to be measured in terms of certain aspects of secondary production, particularly feeding rate, growth efficiency, reproduction and sex ratio.

PART II. FISH

Hydrocarbon levels

The lack of published information on the distribution of hydrocarbons in fish prompted us to initiate an examination of the distribution and amount of alkanes in the flesh and liver tissues of a number of species. A selection of these analyses is given in Table 6. In addition, Mackie, Whittle & Hardy (1974), reported alkane levels for commercially important species from the Firth of Clyde area. The n-alkanes and the branched alkane pristane were present at low concentrations in most of the species examined, particularly in muscle where pristane was not always detected. However, herring (*Clupea harengus*) which can accumulate large quantities of fat in the musculature, e.g. 25 % by weight (Iles & Wood, 1965), also show elevated alkanes (Mackie *et al.*, 1974). Pristane, the predominant constituent in the planktonivorous species, varies considerably as a proportion of the total n-alkanes measured. Whilst the concentrations of n-alkanes in the livers are generally higher than in muscle, a more detailed analysis shows that their distribution has a characteristically high predominance of odd-carbon numbered n-alkanes compared with muscle (Whittle *et al.*, 1974*a*). The predominance is accounted for largely by the higher homologues especially C_{31}. Fig. 5 compares the differing patterns for some species which are typical of those we have analysed from a variety of locations from the Antarctic to Spitzbergen. The origin of these differing

Table 6. *Alkane levels in fish tissues*

Species	Tissue	Hydrocarbons ($\mu g\ g^{-1}$) C_{15}–C_{33}*	Pristane	Reference
Eulachon (*Thaleichthys pacificus*)	Head and body	Not determined	42.0	Ackman, Addison & Eaton (1968)
	Liver	Not determined	54.0	
Mackerel (*Scomber scombrus*)	Muscle	1.9	28.3	Whittle *et al.* (1974*b*)
	Liver	3.2	14.4	
Herring (*Clupea harengus*)	Muscle	11.8†	365.7	Mackie *et al.* (1974)
Plaice (*Pleuronectes platessa*)	Muscle	0.2	ND	Hardy, Whittle, Mackie & Blackman
	Liver	1.1	0.01	(unpublished data)
Gurnard (*Trigla gurnardus*)	Muscle	0.3	ND	Hardy, Whittle, Mackie & Blackman
	Liver	1.1	0.2	(unpublished data)
Bass (*Dicentrarchus labrax*)	Muscle	0.6	ND	Hardy, Whittle, Mackie & Blackman
	Liver	4.8	0.06	(unpublished data)
Whiting (*Gadus merlangus*)	Muscle	0.1	ND	Hardy, Whittle, Mackie & Blackman
	Liver	2.3	0.3	(unpublished data)
Sole (*Solea solea*)	Muscle	0.3	ND	Hardy, Whittle, Mackie & Blackman
	Liver	4.6	0.2	(unpublished data)

ND, not detected.
* C_{15}–C_{33}, the sum of the concentrations of n-alkanes C_{15} to C_{33} inclusive.
† C_{18}–C_{33}.

profiles and the relative abundance of the C_{31} is unclear and it cannot be said with certainty whether the alkanes are largely dietary, metabolic or environmental in origin.

The presence of pristane in fish tissues requires some comment. A major source of pristane in the marine environment is thought to be the metabolism of phytol (Fig. 6) by zooplankton or their gut flora (Avigan & Blumer, 1968). Pristane is also a component of petroleum. If we assume no other sources of this compound, its presence in fish must be dietary or environmental in origin and in either case could suggest a certain stability and lack of further metabolism in the food chain. Blumer *et al.* (1969), have stressed the stability of isoprenoid hydrocarbons in the food chain and Whittle, Mackie, Hardy and McIntyre (1947*b*) showed that in planktonivorous species pristane has a

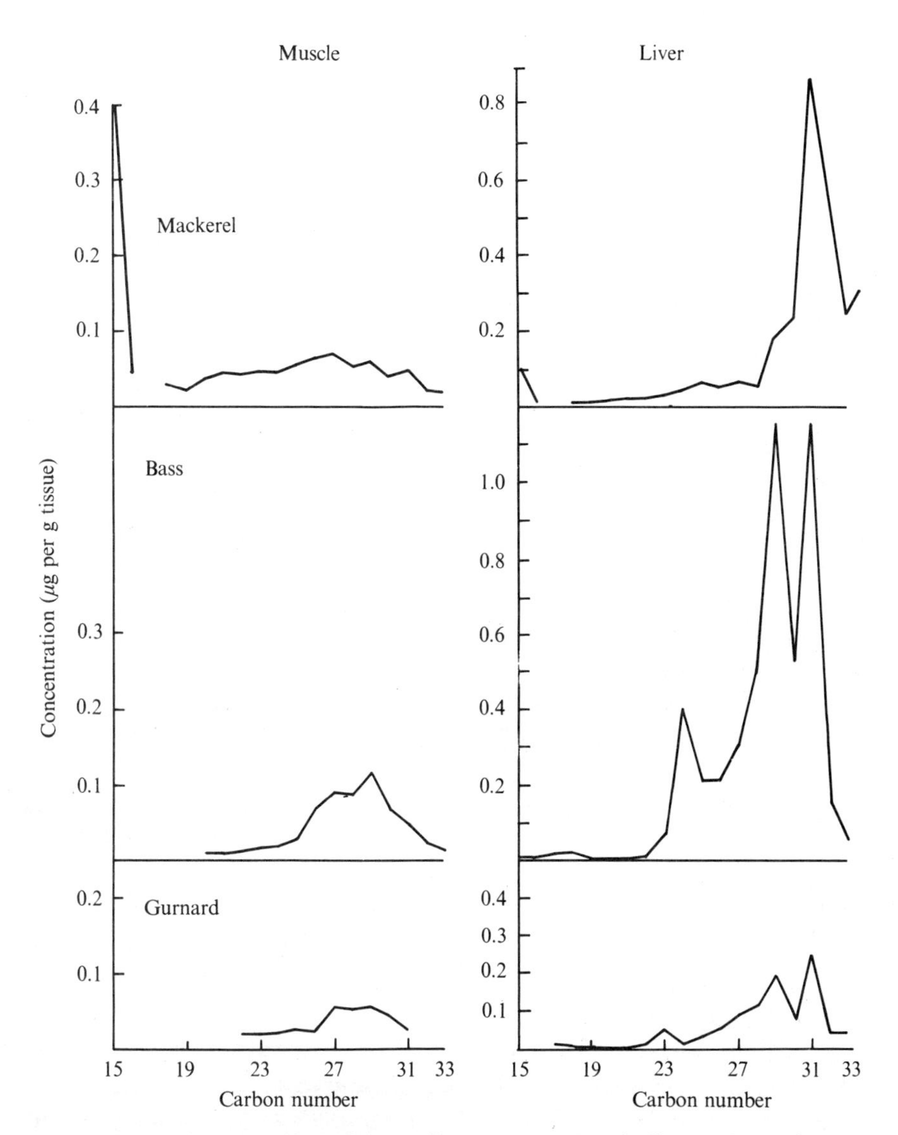

Fig. 5. A comparison of the n-alkane distributions in liver and muscle tissues of fish. The lines between carbon number points aid presentation. They have no mathematical significance.

greater tendency to accumulate than n-alkanes. With respect to the fate of pristane in the food chain it is important to know whether planktonivorous and/or herbivorous fish, or indeed fish in general, have the ability to convert phytol to pristane. The inter-relationships between phytol, pristane, phytane and the various olefins are shown in Fig. 6.

It has been suggested that phytane in the marine environment is

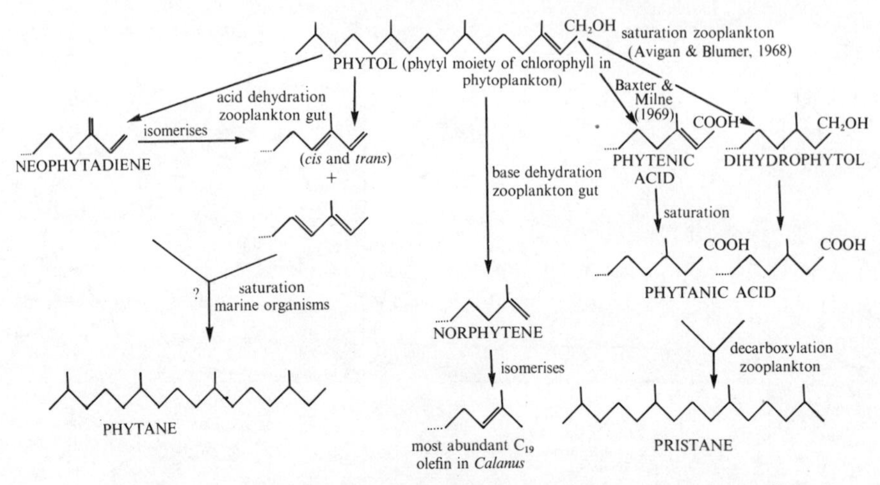

Fig. 6. Inter-relationships between phytol-derived hydrocarbons. Other olefins which have been identified (Blumer *et al.*, 1969) are thought to arise from isoprenoid acids by decarboxylation.

non-biogenic in origin, arising from fossil fuels, and consequently is not detected in samples taken from non-polluted areas. In tissues of fish from Antarctic waters we have tentatively identified phytane at low concentration in livers and flesh (0.001–0.005 $\mu g\ g^{-1}$) which could, in view of the suggestion above be ascribed to contaminated feed or uptake from the environment. However, we might consider whether a minor biochemical pathway could also contribute to its presence. Conceivably this could occur via the saturation of phytadienes (which Blumer, 1967, identified as minor components of the hydrocarbon fraction of basking shark (*Cetorhinus maximus Gunnerus*) liver oil). The phytadienes are probably also present in herring and mackerel (*Scomber scombrus*). After careful and sensitive analysis Blumer was unable to detect phytane in either shark liver oil or zooplankton.

Saturation steps, such as those tentatively suggested for the phytadienes are not common in animals, but there are examples such as the biosynthesis of the polyprenol, dolichol, which has a saturated residue adjacent to the terminal hydroxyl group, and the conversion of phytol to phytanic acid. These mechanisms seem to be highly specific and none is strictly analogous to the proposed saturation of phytadienes. Whether marine vertebrates have this biosynthetic capability is unknown, but in studies of isoprenoid biosynthesis in marine invertebrates Walton & Pennock (1972) have noted dolichol biosynthesis and, more recently, formation of dihydrofarnesol (J. F. Pennock, personal communication).

Detailed analyses of the aromatic hydrocarbon content of fish lipids are not available at present as modern methods are only just beginning to be applied

(Giger & Blumer, 1974). Much of the available information refers to levels of benzo[*a*]pyrene (Mallet, Perdriau & Perdriau, 1963; Bourcart & Mallet, 1965; Scaccini Cicatelli, 1966; Piccinetti, 1968; and Lee *et al.*, 1972).

Toxicity

In general juvenile forms are more vulnerable to the effects of oil than adult fish. Firstly, living near the surface, particularly in the early stages of development, they are exposed to: (i) possible physical contact with the organic-rich surface film which may or may not be heavily contaminated with oil components, and (ii) potentially higher concentrations of soluble, toxic components entering the water column from surface contamination with oil especially when vertical mixing is incomplete. Secondly, their sensitivity to toxic compounds is higher (Moore & Dwyer, 1974).

Crude oils of different origin and petroleum products differ widely in their composition and hence in their contribution to the seawater-soluble fraction, components of which are the most likely toxic agents. In addition, weathering of spilled oil, resulting in the loss of the more volatile and soluble constituents, is known to reduce toxic effects. Moore & Dwyer (1974) in a recent review concluded that the lower-boiling, more soluble aromatic component is consistently implicated as being the primary toxic agent to fish. This is compatible with the conclusion reached in Part I of this chapter that bi- and tri-cyclic aromatic compounds could be responsible for toxicity in crustaceans. Moore & Dwyer (1974) estimated that the soluble aromatic fraction causes toxicity to fish within the range 5–50 ppm and data presented by Anderson *et al.* (1974) for *Cyprinodon variegatus*, *Fundulus similus* and *Menidia beryllina* for two crude oils, a Bunker C oil and No. 2 Fuel Oil, put the toxicity levels at the lower end of this range. In view of the differences between crude oils it is important to study the toxicity of North Sea crude, at least for those species which are important either from an economic point of view or from their position in the food chain. The problems and relevance of toxicity testing have been discussed in detail during a recent Workshop (Beynon & Cowell, 1974).

Behavioural effects

The importance of chemical communication in maintaining the behaviour patterns of fish as well as other marine organisms is well-documented (see review by Hasler, 1970). Animals are sensitive to chemical signals at low concentrations, often well below the ppb range, and may rely extensively on this sensory input to control their attitudes and behaviour (see Todd, Atema & Bardach, 1967). Few investigations have been conducted on the effects of the hydrocarbon components of the water-soluble fraction of crude oil or its products on the ability of fish to perceive such chemical signals at low concentration either by blocking or competing with the sensory 'site' or by providing false signals. Undoubtedly there are technical difficulties in obtaining useful and valid data. However, investigations of this nature should be both stimulating and rewarding since information on even the subtle effects on behaviour is important in assessing, for instance, the ecological effects of oil pollution.

Uptake, release and metabolism (including current experiments)

Availability to fish of hydrocarbons in seawater

In Part I of the chapter the distinction was made between 'dissolved' and 'particulate' hydrocarbons. The accommodation of hydrocarbons in seawater is important when considering the availability of these compounds to fish in their environment. The situation is complex since it is impossible to consider individual compounds or even classes of hydrocarbons alone. Hydrocarbons are in the sea along with a host of other organic compounds which may affect the mechanisms of solution, micellar formation, dispersion, adsorption and absorption. In isolation, the short-chain aliphatics and the low-boiling aromatics are the most soluble. However, Boehm & Quinn (1974) showed that dissolved organic matter in seawater could solubilise n-alkane and isoprenoid hydrocarbons but not aromatic compounds such as phenanthrene and anthracene. Nevertheless, the polynuclear aromatic hydrocarbons (PAHs) can be accommodated in micelles. Boylan & Tripp (1971) observed a group of high-boiling aromatic compounds in seawater extracts of Kuwait oil. Adsorption of hydrocarbons onto particulate matter can also occur.

The major routes by which hydrocarbons present in these various forms are available to fish are either directly via the gills and drinking water or indirectly via the diet. The latter might include organisms which have either adsorbed or absorbed hydrocarbons from the environment. Lee *et al.* (1972) in following the uptake from seawater by fish of dissolved, ^{3}H-labelled benzo[*a*]pyrene, interpreted the results in terms of a positive transfer across the gill surfaces.

The drinking rates of fish depend upon the need to maintain a more or less constant internal osmotic balance. The freshwater teleost drinks little since water enters across the gills and must be excreted copiously in hypotonic urine. The marine teleost drinks heavily to compensate for water lost across the gills and excretes isotonic urine. Elasmobranchs, in which urea is the predominant end-product of nitrogen metabolism and contributes significantly to the osmotic balance of the animal, take in seawater only incidentally with their food. Hence, water balance and thus drinking rate affect the presence in the gut, for instance, of accommodated or particulate hydrocarbons included in the drinking water and the possible absorption of these compounds via the gut.

Tainting

In considering the effects of petroleum or its products on fish and other organisms it should not necessarily be assumed that the hydrocarbons are responsible for the tainting effect. There are few situations where these compounds have been unequivocally implicated in tainting. One of these was the grey mullet (*Mugil japonicus*) tainted in the vicinity of a petroleum refinery complex (Ogata & Miyake, 1970). They concluded that toluene in the effluent was largely responsible for the offensive odour and were able to reproduce the same effect in eels kept in similarly contaminated seawater. In their view, toluene was deposited in the flesh via blood from the branchia, a hypothesis that finds support in the work of Lee *et al.* (1972) on the uptake of naphthalene and benzo[*a*]pyrene from seawater via the gills.

There are many examples of fish becoming tainted in areas heavily contaminated, either consistently or specifically, with petroleum or petroleum products (cf. Connell, 1971). Sidhu, Vale, Shipton & Murray, (1970) found experimentally that mullet (*Mugil cephalus*) became tainted in aquarium water containing 5 ppm kerosene and Deshimaru (1971) noted that yellow tail (*Seriola quinqueradiata*) reared in seawater containing crude oil were tainted after 5 days at 50 ppm or 13 days at 10 ppm. Mann (1969) has shown that the assimilation of tainting compounds is enhanced by the presence of detergents and that phenolic and naphthenic compounds can be responsible for 'oily' taints. Aliphatic and aromatic hydrocarbons can be assimilated into muscle tissues. Thus, in mullet, yellow tail, trout (*Salmo trutta*), Black Sea bream (*Mylio macrocephalus*) and chum salmon (*Oncorhynchus keta*) taint was associated with volatile hydrocarbons in the flesh but the compounds responsible were not unequivocally identified (Shipton, Last, Murray & Vale, 1970; Connell, 1971; Deshimaru, 1971; Mackie, McGill & Hardy, 1972; Motohiro & Inoue, 1973).

Hardy, Mackie, Whittle, Howgate & Eleftheriou (unpublished data) have been following the tainting effects of sand-sunk North Sea crude oil on bottom-dwelling organisms. Preliminary results show that within 1 or 2 days' exposure of plaice (*Pleuronectes platessa*) on a contaminated sandy bottom previously weathered for 24 hours, the fish were heavily tainted at a level corresponding to 200–250 ppm crude oil in the flesh. The tainting substances, which can be recognised by the trained taste panel at threshold concentrations equivalent to 10–30 ppm crude oil in the flesh, are found in aqueous extracts of oil but have somewhat more polar properties by adsorption chromatography than the bulk of the hydrocarbons.

Not all so-called 'petroleum' odours or taints are attributable to petroleum or its products. Dimethyl sulphide formed by thermal decomposition of dimethyl-β-propiothetin, present in phytoplankton, has been identified as a 'petroleum' taint (Motohiro, 1962; Ackman, Hingley & May, 1967).

Studies of the hydrocarbons of the natural diet

The source of hydrocarbons in fish may be the sea, the diet or intrinsic metabolism. Blumer (1967) concluded that the zamenes, phytadienes and pristane in the hydrocarbon fraction of basking shark liver oil lipids were derived virtually intact from the lipids of the zooplankton on which the shark had fed. On the basis of the hydrocarbon pattern he considered the source to have been *Calanus* sp. acquired in colder waters some distance from where the fish was caught. Blumer *et al.* (1969) reported that the livers of herring from the Gulf of Maine had a hydrocarbon distribution of the same type as *Calanus*, the principal species in that region. New Jersey herring, however, had a very different composition which they suggested was due to differences in their principal food source. Mackie *et al.* (1974) commented on the close similarity between the pattern of the alkane fraction of herring flesh and that of the mixed plankton sample taken in the same area in which the herring were caught and upon which they were presumed to have been feeding. This observation was taken a stage further by Whittle *et al.* (1974*b*) in a similar comparison of plankton (in which *Calanus finmarchicus* was abundant), sprat (*Sprattus sprattus*) and mackerel taken from the same location and forming a clear food chain, judging from an examination of stomach contents. The alkane fraction of the mackerel flesh was similar to that of whole sprat and the plankton suggesting that, as in the other examples, the hydrocarbons were being assimilated at each stage virtually intact. On the other hand the mackerel livers, unlike the herring livers examined by Blumer *et al.* (1969), showed the now familiar odd-carbon predominance of n-alkanes routinely found in fish livers described earlier. The close similarity between the dietary hydrocarbons and those in the tissues of the planktonivorous species described above, particularly in the muscle where lipids are deposited, suggests that the hydrocarbons are deposited with the lipids.

Since the dietary hydrocarbon intake is likely to vary considerably in both range and composition with the species in the diet, and since the results above show a close parallel between diet and tissue hydrocarbons, it seems that the turnover rate of hydrocarbons in these tissues must be more rapid than hitherto suspected. This possibility finds support in the deposition of hydrocarbons in the flesh of fish exposed to water contaminated with petroleum or petroleum products as described earlier. There is a rapid acquisition of the taint and the muscle hydrocarbon patterns are akin to those of the contaminating oil. In addition the examples (chum salmon, mullet, trout and yellow tail) on which these observations have been made also deposit lipids in the muscle tissue, providing further support for the suggestion above that hydrocarbons are deposited with the lipids.

Studies with oil in the diet

Observations of ingested oil in nature are rare. Horn, Teal & Backus (1970) noted the presence of oil in the stomachs of sauries (*Scomberesox saurus*) but no analyses of the tissues were carried out. The fate of oil experimentally incorporated in the diet of fish has been studied in yellow tail by Deshimaru (1971), in plaice by Blackman & Mackie (1973) and in codling (*Gadus morrhua*) by Hardy, Mackie, Whittle & McIntyre (1974).

A diet containing about 1 % crude oil was fed to yellow tail for 13 days. After the eighth day a slight oily taste at a fairly constant level was perceptible in the flesh. Alkanes and other oil-derived components were detected in the head space vapours.

The plaice were fed for 10 days with shrimps each containing 10 μl of topped Kuwait crude oil and subsequently maintained without oil for a further 30 days. During oil-feeding, n-alkanes were deposited in the flesh and also appeared in the faeces. The faecal n-alkane pattern closely resembled the n-alkane distribution in the oil which they had been fed and this component must be presumed to have been essentially unchanged and probably unabsorbed. The uptake and elimination of n-alkanes (C_{15}–C_{22}) derived from the oil was followed in the muscle. They were identified in the flesh within 5 days but less than 2 % of the alkanes fed appeared unchanged in the muscle and they quickly disappeared to reach background levels once the feeding of oil ceased. Plaice store significant amounts of lipid in the muscle under the skin in the region of the lateral fins. It is not known whether the n-alkanes were deposited in this region but the results appear to support the suggestion that dietary hydrocarbons are located in the lipid deposits.

The codling were fed a squid diet containing 1 mg of the same topped Kuwait crude oil (5.7 % n-alkanes C_{15}–C_{33}) per day for 175 days and then maintained for a further 109 days without oil. No deposition of n-alkanes

Table 7. *A comparison of alkane contents for oil-fed and control codling (μg/g wet wt)*

	Oil-feeding period (days)							
	0				62			
	Liver		Muscle		Liver		Muscle	
	C_{15}–C_{33}*	PR	C_{15}–C_{33}	PR	C_{15}–C_{33}	PR	C_{15}–C_{33}	PR
Control	3.37[a]	0.19	0.23	0.01	3.37[a]	0.19	0.23	0.01
Treated	—	—	—	—	21.37[a]	0.16	0.21	0.02
	Oil feeding period (days)							
	126				175			
	Liver		Muscle		Liver		Muscle	
	C_{15}–C_{33}	PR	C_{15}–C_{33}	PR	C_{15}–C_{33}	PR	C_{15}–C_{33}	PR
Control	2.56[a]	0.34	0.18	0.05	4.91[a]	0.15	0.15	0.01
Treated	13.70[a]	0.36	0.21	0.03	29.9[a]	0.36	0.21	0.01
	Recovery period (days)							
	237				284			
	Liver		Muscle		Liver		Muscle	
	C_{15}–C_{33}	PR	C_{15}–C_{33}	PR	C_{15}–C_{33}	PR	C_{15}–C_{33}	PR
Control	7.82[a]	0.29	0.20	0.01	3.41[b]	0.14	0.12	0.01
Treated	16.12[a]	0.23	0.18	0.02	13.30[c]	0.96	0.20	0.03

* C_{15}–C_{33}, the sum of the concentrations of n-alkanes C_{15} to C_{33} inclusive.
† PR, pristane.
[a] Mean of three fish; [b] Mean of four fish; [c] mean of five fish.

(C_{15}–C_{33}) occurred in the muscle. However, during oil-feeding the level in the liver increased by a factor of 6, remained high until oil-feeding stopped and then fell to about 4 times the previous background level by the end of the experiment (Table 7). The characteristic odd-carbon predominance noted earlier in fish livers was masked completely by the deposited alkanes and the liver content assumed a smooth profile more characteristic of muscle analyses (Fig. 5). After the oil was withdrawn the original profile was not regained by the end of the experiment. The pattern of the liver alkanes differed markedly from that of the oil as Blackman & Mackie (1973) had found in the plaice

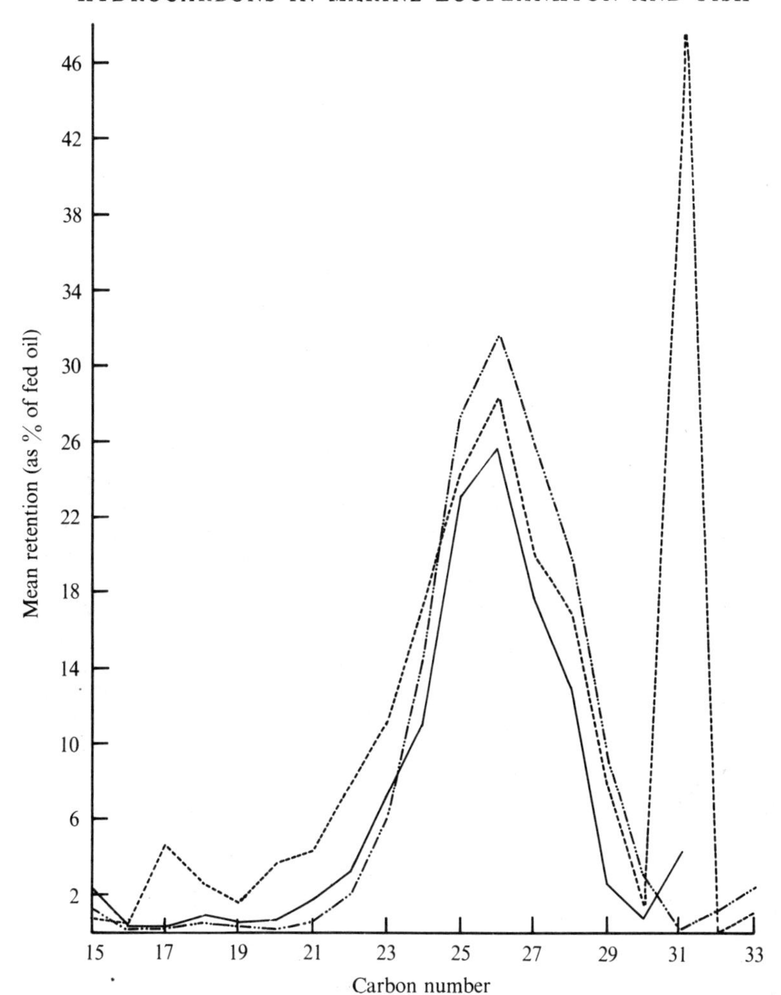

Fig. 7. Percentage retention of n-alkanes in the livers of oil-fed fish. Groups of three fish were fed for: – – –, 62 days; ——, 126 days; and 175 days. Each curve represents a mean of three fish.

muscle. When the liver alkanes were examined individually on a percentage retention basis an interesting pattern emerged (Fig. 7), there being a continuous variation in the retention of alkanes through the homologous series. Retention peaked with C_{26}, being some 100-fold higher than the short-chain homologues, and fell again towards the longer-chain homologues.

Further examination of the results (Hardy *et al.*, unpublished data) has shown that the daily dose of oil contained about 1 μg of pristane, but the amount of pristane recovered in the tissues showed such a variation that no similar trends to the n-alkanes were immediately obvious. On a wet tissue weight basis pristane levels throughout the experiment were about an order of magnitude higher in liver than in muscle (Table 7). Calculation of the loss of n-alkanes from the liver following withdrawal of the oil has shown that heptadecane as well as homologues larger than C_{28} tended to be retained. The remainder tended to be lost, the variation between homologues being about a factor of 2.

Cod seldom have a muscle lipid content higher than 1 %, a value lower than that of the previously mentioned species (Dambergs, 1964; Love, 1970, p. 228). The major site of lipid deposition is the liver which may contain some 60 % fat and occupy 7 % of the body weight. This marked centralisation of fat deposition in the liver may well account for the lack of any apparent effect of oil-feeding in the muscle tissues of cod compared with the marked changes noted in the muscular tissues of other species.

The differences in retention of the individual n-alkanes could arise from a number of possible mechanisms, as Hardy *et al.* (1974) pointed out. Discrimination in assimilation may not be confined to cod and may have implications for experiments in which fish are fed individual hydrocarbons.

Studies with individual hydrocarbons

Few studies with individual hydrocarbons have been conducted with fish, most of these being concerned with the carcinogen benzo[*a*]pyrene, either added to the aquarium water (Lee *et al.*, 1972) or to the diet (Scaccini, Scaccini Cicatelli, Marani & Leonardi, 1970). In addition Lee's group used naphthalene. R. J. Pentreath (personal communication) fed plaice with benzo[*a*]pyrene and hexadecane and also added these hydrocarbons to the aquarium water. Whittle *et al.* (1974*b*) briefly referred to the feeding of hexadecane to codling.

Lee *et al.* (1972) followed the uptake, discharge and metabolism of [^{14}C]naphthalene and [^{3}H]benzo[*a*]pyrene in seawater with the sand goby (*Gillichthys mirabilis*), sculpin (*Oligocottus maculosus*) and sand dab (*Citharichthys stignaeus*). All three species took up hydrocarbons through the gills and activity then built up in the liver. There naphthalene, which apparently can be dealt with in larger quantities than benzo[*a*]pyrene, was metabolised in part to 1,2-dihydro-1,2-dihydroxynaphthalene after 24 hours' exposure whilst 7,8-dihydro-7,8-dihydroxybenzo[*a*]pyrene was tentatively identified as the major product of benzo[*a*]pyrene metabolism. Subsequently the hydrocarbons and their metabolites were transported and stored in the gall bladder. Hydrocarbons deposited in the flesh tended to be lost on transfer to clean water, naphthalene loss occurring at a faster rate than that of benzo[*a*]-pyrene. Some water-soluble metabolites, probably in conjugated form,

Table 8. *Distribution of ^{14}C activity (as % of total activity found) at intervals after feeding squid containing [^{14}C]benzo[a]pyrene together with [^{14}C]hexadecane to codling*

Fraction	% activity recovered after		
	48 h	72 h	96 h
Stomach	83.5	32.9	37.1
Liver	2.2	3.6	3.4
Gall fluid	0.6	12.5	1.9
Intestinal contents	12.9	8.6*	36.3
Urine	0	Trace	0.03
Aquarium residue (mainly faeces)	0.5	41.4	20.7
Aquarium water	0	0.2	0.12
Plasma†	0.03	0.08	0.05
Blood†	0	0.06	0.04
Gills	0.2	0.5	0.4
Spleen	0.06	0.1	0.1

* A loss of at least 50 % occurred during dissection.
† Expressed on a per gram basis.

were excreted via the urine. Lee's fish were not fed and probably tended to accumulate bile in the gall bladder (Love, 1970, p. 223). Thus storage in the gall bladder may have been an artefact of the experimental design. Menger, Rhee & Mandell (1973) reported that naphthalene could be bound in the core of bile salt micelles near the C_{19} angular methyl groups. This additional, possible mechanism of 'detoxifying' naphthalene could account for the observation that the fish livers apparently had a greater capacity to deal with naphthalene.

The metabolites mentioned above closely follow the routes of metabolism of PAHs described in mammals (reviewed by Corner, 1975) and also outlined for naphthalene metabolism in the marine crab *Maia squinado* (Corner *et al.*, 1973). Simpson & Youngson (1974) with rainbow trout (*Salmo gairdneri*) propose that hydrocarbons can be transported to the other body tissues in the blood plasma mainly in association with plasma lipoproteins. However, the amounts injected intravenously were well in excess of the levels which might be expected to occur naturally.

Lee *et al.* (1972) suggested that the process of transfer and retention of hydrocarbons would be likely to be different depending on whether it occurs in the gut or gills. Scaccini *et al.* (1970) reported on feeding a benzo[*a*]pyrene-contaminated diet to goldfish (*Carassius auratus*) over a range of values from 0.7 μg to 50 μg hydrocarbon per fish per day. They concluded that once

Table 9. *Distribution of* ^{14}C *activity (as % of total activity recovered in the lipid fraction) in various tissues of codling* 96 *hours after feeding* [^{14}C]*hexadecane or* [^{14}C]*benzo*[a]*pyrene*

Tissue	% activity recovered in lipid fraction	
	Hexadecane	Benzo[*a*]pyrene
Stomach	0.8	0.3
Stomach contents	86.2	41.7
Pyloric caecae	1.1	1.2
Liver	0.5	0.4
Gall fluid	0.5	26.3
Intestinal wall	0.2	0.3
Intestinal contents	2.3	7.2
Urine	0	0
Aquarium residue (mainly faeces)	0.9	0.6
Aquarium water	7.3	22.0
Gills	0.2	0
Muscle	0	0

absorbed via the gut it is deposited principally in the liver and secondly in the gonads and is probably associated with the high lipid content of these tissues. The hydrocarbon was also deposited in muscle and a fraction was eliminated unchanged through the kidney.

We have carried out some preliminary studies (Hardy, Whittle, Mackie & Murray, unpublished data) by feeding single doses of [^{14}C]benzo[*a*]pyrene and [^{14}C]hexadecane to codling and examining the distribution of activity in various tissues and organs. Initially, the two compounds were fed together and the activity followed over 96 hours (Table 8). In keeping with the high level of activity found in the stomach, the squid food in which the hydrocarbons were incorporated was still recognisable there after 96 hours and its slow digestion could well be related to the general stress condition of aquarium-held fish. Some material was apparently unabsorbed and excreted with the faeces whilst that which was absorbed primarily followed the route: liver, bile, gall bladder and faeces to the exterior. Little activity was recovered in the blood or plasma and consequently in the flesh or urine. The form of the recovered activity was investigated by feeding codling either benzo[*a*]pyrene or hexadecane and examining them after 96 hours (Table 9).

As in the previous experiment a high proportion of the activity remained associated with the stomach contents. The greater recovery of activity from areas other than the stomach in the benzo[*a*]pyrene-fed fish does not necessarily mean that this compound was handled more quickly; it could be explained simply by variation in individual digestion rates. In terms of lipid fraction activity released from the stomach, roughly 10 times as much accumulated in the gall bladder bile in the benzo[*a*]pyrene-fed fish compared with those fed hexadecane. In the latter, a greater proportion of the activity had been excreted or was *en route* to excretion. No lipid fraction activity was recovered from muscle tissue with either compound. In the hexadecane-fed fish much of the activity, notably in the intestinal wall as well as in the intestinal and stomach contents, was associated with the residues after lipid extraction. The recovered activity from stomach contents contained a more polar product of benzo[*a*]pyrene which appeared to be analogous to material produced by simple acid treatment of benzo[*a*]pyrene. Acid conditions in the stomach might account for this effect, though alternatively it is conceivable that the polar products could have been regurgitated into the stomach from the upper gut.

General conclusions

In fish the concentration of n-alkanes is generally higher in liver than in muscle tissue. The highest tissue levels are found in planktonivorous species which all have a relatively lipid-rich and hydrocarbon-rich diet and can accumulate high concentrations of fat in the muscle. There is no evidence to suggest that the n-alkanes accumulate in fish tissues, but the same may not be true for the branched alkanes such as pristane which seems to be more stable in the food chain. More analytical data needs to be obtained to assess the presence and distribution of aromatic hydrocarbons in fish tissues.

In planktonivorous species, the clear similarity between the dietary alkanes and those of the muscle or liver tissues and the implied high turnover rate of the compounds in the tissues, strongly suggest that intrinsic metabolism is a minor source of alkanes in these fish. The major contribution is the diet which may include non-biogenic components. A further contribution can be made directly from the environment either by absorption across the gills or from the water that is drunk. At present, however, such alkanes can only be identified with certainty when hydrocarbon pollution is high.

The patterns of n-alkanes in the liver and muscle tissues of the planktonivorous species are broadly similar to those in the same tissues of all the other fish species analysed so far, including ones which have highly centralised lipid reserves in the liver and store little in the muscle. They all show the smooth profile in muscle tissue and the odd-carbon predominance in liver tissue. This similarity suggests that intrinsic metabolism may also be a minor source of n-alkanes in the non-planktonivorous species.

When the dietary intake or the environmental load of alkanes is high they tend to be deposited in tissues where lipid is deposited. This is demonstrated amply by analyses of the planktonivorous species, by the examples of tainting and by the various feeding experiments, especially those on cod. No deposition occurred in the muscle tissue, where lipid deposition, if any, is very low.

In the feeding experiments where the input was known and quantified it was found that the n-alkanes were not assimilated in the proportions present in the feed, some homologues being retained to a greater extent than others. Insufficient experimental work has been carried out to determine whether this phenomenon of apparent discrimination is generally applicable in fish. For instance, results to date do not cover fatty species or an examination of a range of aromatic compounds. The work emphasises the dangers in extrapolating results obtained with a single compound to predict the behaviour of the entire class or group to which it belongs.

Our knowledge of the assimilation and metabolism of aromatic hydrocarbons in fish rests largely on experience with the PAH, benzo[*a*]pyrene. Once absorbed it is hydroxylated in the liver, concentrated in the bile, at least partially in conjugated form, and released into the gut as the gall bladder empties in response to food. A similar pathway seems to exist for naphthalene.

Tainting of fish in the presence of high environmental concentrations of petroleum or petroleum products is associated with the presence in the tainted flesh of hydrocarbons, including alkanes, derived from these sources. Only in one or two of the examples quoted have the substances responsible for the taint been unequivocally identified as hydrocarbons and it is likely that more polar compounds are implicated. If, under the same conditions of exposure, the tainting compounds are assimilated and deposited in an analogous way to the n-alkanes, it should be possible to predict which fish are likely to be tainted and to what extent.

The bi- and tri-cyclic aromatic hydrocarbons are among those components of oil which are toxic to fish. However, the concentrations at which toxicity is likely to occur under natural conditions are not readily derived from the standard toxicity tests which have been carried out on oils and oil fractions. Nevertheless, the methods are considered to be sufficiently sensitive to rank the toxicities of the test substances. As in many bioassays, a major problem is the difficulty of devising test conditions which are either analogous to or can be extrapolated to the natural situation.

We are grateful to Howard Platt of the British Antarctic Survey for the supply of, and assistance in the analysis of, the Antarctic fish described in Part II of this paper.

References

Ackman, R. G., Addison, R. F. & Eaton, C. A. (1968). Unusual occurrence of squalene in a fish, the Eulachon (*Thaleichthys pacificus*). *Nature, London*, **220**, 1033–4.

Ackman, R. G., Hingley, J. & May, A. W. (1967). Dimethyl-β-propiothetin and dimethyl sulphide in Labrador cod. *Journal of the Fisheries Research Board of Canada*, **24**, 457–61.

Andelman, J. B. & Suess, M. J. (1970). Polynuclear aromatic hydrocarbons in the water environment. *Bulletin of the World Health Organization*, **43**, 479–508.

Anderson, J. W., Neff, J. M., Cox, B. A., Tatem, H. E. & Hightower, G. M. (1974). Characteristics of dispersions and water soluble extracts of crude and refined oils and their toxicity to estuarine crustaceans and fish. *Marine Biology*, **27**, 75–88.

Avigan, J. & Blumer, M. (1968). On the origin of pristane in marine organisms. *Journal of Lipid Research*, **9**, 350–2.

Barbier, M., Joly, D., Saliot, A. & Tourres, D. (1973). Hydrocarbons from sea water. *Deep-Sea Research*, **20**, 305–14.

Barnett, C. J. & Kontogiannis, J. E. (1975). The effect of crude oil fractions on the survival of a tidepool copepod, *Tigriopus californicus*. *Environmental Pollution*, **8**, 45–54.

Baxter, J. H. & Milne, G. W. A. (1969). Phytenic acid: identification of five isomers in chemical and biological products of phytol. *Biochimica et Biophysica Acta*, **176**, 265–77.

Beynon, L. R. & Cowell, E. B. (1974). *Ecological aspects of toxicity testing of oils and dispersants*. Applied Science Publishers, Barking, Essex.

Blackman, R. A. A. & Mackie, P. R. (1973). Preliminary results of an experiment to measure the uptake of *n*-alkane hydrocarbons by fish. ICES, CM 1973/E:23 Fisheries Improvement Committee, Lisbon.

Blumer, M. (1967). Hydrocarbons in digestive tract and liver of a Basking Shark. *Science*, **156**, 390–1.

Blumer, M., Guillard, R. R. L. & Chase, Y. (1971). Hydrocarbons of marine phytoplankton. *Marine Biology*, **8**, 183–9.

Blumer, M., Mullin, M. M. & Thomas, D. W. (1970). Pristane in the marine environment. *Helgoländer Wissenschaftliche Meeresuntersuchungen*, **10**, 187–201.

Blumer, M., Robertson, J. C., Gordon, J. E. & Sass, J. (1969). Phytol-derived C_{19} di- and triolefinic hydrocarbons in marine zooplankton and fishes. *Biochemistry*, **8**, 4067–74.

Boehm, P. D. & Quinn, J. G. (1974). The solubility behaviour of No. 2 fuel oil in sea water. *Marine Pollution Bulletin*, **5**, 101–4.

Booth, J. & Sims, P. (1974). 8,9-Dihydro-8,9-dihydroxybenz[*a*]anthracene-10,11-oxide: a new type of polycyclic aromatic hydrocarbon metabolite. *FEBS Letters*, **47**, 30–3.

Bourcart, J. & Mallet, L. (1965). Pollution marine des rives de la région centrale de la mer Tyrrhèniene (baie de Naples) par les hydrocarbures polybenzéniques du type benzo-3,4 pyréne. *Comptes Rendus hebdomadaires des Séances de l'Academie des Sciences*, Paris, **260**, 3729–34.

Boylan, D. B. & Tripp, B. W. (1971). Determination of hydrocarbons in sea water extracts of crude oil and crude oil fractions. *Nature, London*, **230**, 44–7.

Burwood, R. & Spears, G. C. (1974). Photo-oxidation as a factor in the environmental dispersal of crude oil. *Estuarine and Coastal and Marine Science*, **2**, 117–35.

Connell, D. W. (1971). Kerosene-like tainting in Australian mullet. *Marine Pollution Bulletin*, **2**, 188–9.

Conover, R. J. (1971). Some relations between zooplankton and bunker C oil in Chedabucto Bay following the wreck of the tanker Arrow. *Journal of the Fisheries Research Board of Canada*, **28**, 1327–30.

Copin, G. & Barbier, M. (1971). Substances organiques dissoutes dans l'eau de mer. Premiers résultats de leur fractionnement. *Cahiers océanographiques* **23**, 455–64.

Corner, E. D. S. (1975). The fate of fossil fuel hydrocarbons in marine animals. *Proceedings of the Royal Society of London*, Ser. B, **189**, 391–413.

Corner, E. D. S., Head, R. N. & Kilvington, C. C. (1972). On the nutrition and metabolism of zooplankton. VIII. The grazing of *Biddulphia* cells by *Calanus helgolandicus*. *Journal of the Marine Biological Association of the United Kingdom*, **52**, 847–61.

Corner, E. D. S., Head, R. N., Kilvington, C. C. & Marshall, S. M. (1974). On the nutrition and metabolism of zooplankton. IX. Studies relating to the nutrition of over-wintering *Calanus*. *Journal of the Marine Biological Association of the United Kingdom*, **54**, 319–31.

Corner, E. D. S., Kilvington, C. C. & O'Hara, S. C. M. (1973). Qualitative studies on the metabolism of naphthalene in *Maia squinado* (Herbst). *Journal of the Marine Biological Association of the United Kingdom*, **53**, 819–32.

Dambergs, N. (1964). Extractives of fish muscle. 4. Seasonal variations of fat, water-solubles, protein and water in cod (*Gadus morhua* L.) fillets. *Journal of the Fisheries Research Board of Canada*, **21**, 703–9.

Deshimaru, O. (1971). Studies on the pollution of fish meat by mineral oils. I. Deposition of crude oils in fish meat and its detection. *Bulletin of the Japanese Society of Scientific Fisheries*, **37**, 297–301.

Dunning, A. & Major, C. W. (1974). The effect of cold sea water extracts of oil fractions upon the blue mussel, *Mytilus edulis*. In *Pollution and physiology of marine organisms* (ed. F. J. Vernberg & W. B. Vernberg), pp. 349–62. Academic Press, New York & London.

Ekwall, P. & Sjøblom, L. (1952). Butyric acid and lactic acid in aqueous solutions as solubilizers for carcinogenic hydrocarbons. *Acta Chemica Scandinavica*, **6**, 96–100.

Elmamlouk, T. H., Gessner, T. & Brownie, A. C. (1974). Occurrence of cytochrome P-450 in hepatopancreas of *Homarus americanus*. *Comparative Biochemistry and Physiology*, **48B**, 419–25.

Frankenfeld, J. W. (1973). Factors governing the fate of oil at sea; variations in the amounts and types of dissolved or dispersed materials during the weathering process. In *Proceedings of the Joint Conference on Prevention and Control of Oil Spills*, pp. 485–95. American Petroleum Institute, New York.

Freegarde, M., Hatchard, C. G. & Parker, C. A. (1971). Oil spill at sea; its identification, determination and ultimate fate. *Laboratory Practice*, **20**, 35–40.

Giger, W. & Blumer, M. (1974). Polycyclic aromatic hydrocarbons in the environment. Isolation and characterization by chromatography, visible, ultraviolet, and mass spectrometry. *Analytical Chemistry*, **46**, 1663.

Gordon, D. C., Keizer, P. D. & Dale, J. (1974). Estimates using fluorescence spectroscopy of the present state of petroleum hydrocarbon contamination in the water column of the north west Atlantic ocean. *Marine Chemistry*, **2**, 251–61.

Hardy, R., Mackie, P. R., Whittle, K. J. & McIntyre, A. D. (1974). Discrimination in the assimilation of *n*-alkanes in fish. *Nature, London*, **252**, 577–8.

Hasler, A. D. (1970). Chemical ecology of fish. In *Chemical ecology* (ed. E. Sondheimer & J. B. Simeone), Academic Press, New York & London.

Holder, G., Yagi, H., Dansette, P., Jerina, D. M., Levin, W., Lu, A. Y. & Conney, A. H. (1974). Effects of inducers and epoxide hydrase on the metabolism of benzo[*a*]pyrene by liver microsomes and a reconstituted system: analysis by high pressure liquid chromatography. *Proceedings of the National Academy of Sciences, USA*, **71**, 4356–60.

Horn, M. H., Teal, J. M. & Backus, R. H. (1970). Petroleum lumps on the surface of the sea. *Science*, **168**, 245–6.

Iles, T. D. & Wood, R. J. (1965). The fat/water relationship in North Sea herring (*Clupea harengus*), and its possible significance. *Journal of the Marine Biological Association of the United Kingdom*, **45**, 353.

Jeffrey, L M. (1970). Lipids of marine waters. In *Symposium on Organic Matter in Natural Waters*, (ed. D. W. Wood), pp. 55–76. University of Alaska Institute of Marine Science Occasional Papers No. 1.

Keizer, P. D. & Gordon, D. C. Jr (1973). Detection of trace amounts of oil in sea water by fluorescence spectroscopy. *Journal of the Fisheries Research Board of Canada*, **30**, 1039–46.

Lee, R. F. (1975). Fate of petroleum hydrocarbons in marine zooplankton. In *Proceedings of the 1975 Conference on Prevention and Control of Oil Spills*. American Petroleum Institute, New York.

Lee, R. F., Sauerheber, R. & Benson, A. A. (1972). Petroleum hydrocarbons: uptake and discharge by the marine mussel, *Mytilus edulis*. *Science*, **177**, 344–6.

Lee, R. F., Sauerheber, R. & Dobbs, G. H. (1972). Uptake, metabolism and discharge of polycyclic aromatic hydrocarbons by marine fish. *Marine Biology*, **17**, 201–8.

Levy, E. M. (1971). The presence of petroleum residues off the east coast of Nova Scotia, in the Gulf of St Lawrence, and the St Lawrence River. *Water Research*, **6**, 723–33.

Love, R. M. (1970). Depletion. In *Chemical biology of fishes* (ed. R. M. Love), Academic Press, New York & London.

Lu, A. Y. H., Kuntzmann, R., West, S., Jacobson, M. & Conney, A. H. (1972). Reconstituted liver microsomal enzyme system that hydroxylates drugs, other foreign compounds and endogenous substrates – II. *Journal of Biological Chemistry*, **247**, 1727–34.

Mackie, P. R., McGill, A. S. & Hardy, R. (1972). Diesel oil contamination of Brown Trout (*Salmo trutta* L.). *Environmental Pollution*, **3**, 9–16.

Mackie, P. R., Whittle, K. J. & Hardy, R. (1974). Hydrocarbons in the marine environment. I. *n*-alkanes in the Firth of Clyde. *Estuarine and Coastal Marine Science*, **2**, 359–74.

Mallet, L. & Lami, R. (1964). Recherche sur la pollution du plancton par les hydrocarbures polybenzénique du type benzo-3,4-pyréne dans l'estuaire de la Rance. *Comptes Rendus des Séances de la Société de Biologie*, **158**, 2261–2.

Mallet, L., Perdriau, L. V. & Perdriau, J. (1963). Pollution par les hydrocarbures polybenzéniques du type benzo-3,4-pyréne de la région occidentale de l'océan glacial Arctique. *Comptes Rendus hebdomadaires des Séances de l'Académie des Sciences*, Paris, **256**, 3487–9.

Mann, H. (1969). Factors affecting the taste of fish. *Fette Seifen Anstrmittel*, **71**, 1021–4.

Menger, F. M., Rhee, J. H. & Mandell, L. (1973). Binding site of naphthalene to bile salt micelles as determined by ^{1}H nuclear magnetic resonance. *Chemical Communications*, **23**, 918–19.

Mileikovsky, S. A. (1970). The influence of pollution on pelagic larvae of bottom invertebrates in marine nearshore and estuarine waters. *Marine Biology*, **6**, 350–6.

Mironov, O. G. (1969). Effect of oil pollution upon some representatives of Black Sea zooplankton. *Zoologicheskii Zhurnal*, **48**, 980–4.

Moore, S. F. & Dwyer, R. L. (1974). Effects of oil on marine organisms: a critical assessment of published data. *Water Research*, **8**, 819–27.

Morello, A., Bleeker, W. & Agosin, M. (1971). Cytochrome P-450 and hydroxylating activity of microsomal preparation from whole houseflies. *Biochemical Journal*, **124**, 199–205.

Morris, R. J. (1974). Lipid composition of surface films and zooplankton from the Eastern Mediterranean. *Marine Pollution Bulletin*, **5**, 105–8.

Motohiro, T. (1962). Studies on the petroleum odour in canned chum salmon. *Bulletin of the Faculty of Fisheries, Hokkaido University*, **10**, 2–65.

Motohiro, T. & Inoue, N. (1973). *n*-Paraffins in polluted fish by crude oil from 'Juliana' wreck. *Bulletin of the Faculty of Fisheries, Hokkaido University*, **23**, 204–8.

Ogata, M. & Miyake, Y. (1970). Offensive odour substance in the evil-smelling fish from the sea polluted by petroleum and petrochemical industrial waste. Report 1. Identification of offensive odour substance. *Acta Medica, Okayama*, **21**, 471–81.

Piccinetti, C. (1968). Diffusione dell'idrocarburo cancerigeno benzo-3,4 pirene nell'alto e medio Adriatico. *Archivo di Oceanografia e Limnologia*, **15**, Suppl., 169–83.

Scaccini Cicatelli, M. (1966). Il benzo-3,4 pirene, drocarburo cancerigeno, nell'ambiente marino. *Archivio Zoologico Italiano*, **51**, 747–74.

Scaccini, A., Scaccini Cicatelli, M., Marani, F. & Leonardi, V. (1970). Dosaggi spettrofotofluorimetrici ed osservazioni mediante microscopia a fluorescenza del benzo-3,4 pirene su organi e tessuti di *Carassius auratus*. *Note del Laboratorio di Biologia Marina e Pesca-Fano Annesso All'Istituto Zoologico Dell'Universita di Bologna*, **3** (6), 105–44.

Scheur, P. J. (1973). *Chemistry of marine natural products*. Academic Press, New York & London.

Shipton, J., Last, J. H., Murray, K. E. & Vale, G. L. (1970). Studies on a kerosene-like taint in mullet (*Mugil cephalus*). II. Chemical nature of the volatile constituents. *Journal of the Science of Food and Agriculture*, **21**, 433–6.

Sidhu, G. S., Vale, G. L., Shipton, J. & Murray, K. E. (1970). A kerosene-like taint in mullet (*Mugil cephalus*). In *Marine pollution and sea life*, ed. M. Ruivo, pp. 546–50. Fishing News (Books) Ltd, London.

Simpson, T. H. & Youngson, A. F. (1974). The transport and mobilization of fat soluble pollutants in the rainbow trout: the effect of changes in physiological status. ICES CM 1974/M:18. Anadromous and Catadromous Committee, Copenhagen.

Spooner, M. F. & Corkett, C. J. (1974). A method for testing the toxicity of suspended oil droplets on planktonic copepods used at Plymouth. In *Ecological aspects of toxicity testing of oils and dispersants* (ed. L. R. Beynon & E. B. Cowell), pp. 69–74. Applied Science Publishers, Barking, Essex.

Stegeman, J. J. & Teal, J. M. (1973). Accumulation, release and retention of petroleum hydrocarbons by the oyster *Crassostrea virginica*. *Marine Biology*, **22**, 37–44.

Terrière, L. C., Boose, R. B. & Roubal, W. T. (1961). The metabolism of naphthalene and 1-naphthol by houseflies and rats. *Biochemical Journal*, **79**, 620–3.

Todd, J. H., Atema, J. & Bardach, J. E. (1967). Chemical communication in social behaviour of a fish, the yellow bullhead (*Ictalurus natalis*). *Science*, **158**, 672–3.

Walton, M. J. & Pennock, J. F. (1972). Some studies on the biosynthesis of ubiquinone, isoprenoid alcohols, squalene and sterols by marine invertebrates. *Biochemical Journal*, **127**, 471–9.

Whitlock, J. P Jr & Gelboin, H. V. (1974). Aryl hydrocarbon (benzo[*a*]pyrene) hydroxylase induction in rat liver cells in culture. *Journal of Biological Chemistry*, **249**, 2616–23.

Whittle, K. J., Mackie, P. R. & Hardy, R. (1974*a*). Hydrocarbons in the marine eco-system. *South African Journal of Science*, **70**, 141–4.

Whittle, K. J., Mackie, P. R., Hardy, R. & McIntyre, A. D. (1973). A survey of hydrocarbons in Scottish coastal waters. ICES CM 1973/E:24. Fisheries Improvement Committee, Lisbon.

Whittle, K. J., Mackie, P. R., Hardy, R. & McIntyre, A. D. (1974*b*). The fate of *n*-alkanes in marine organisms. ICES CM 1974/E:33. Fisheries Improvement Committee, Copenhagen.

Zobell, C. E. (1971). Sources and biodegradation of carcinogenic hydrocarbons. In *Proceedings of the Joint Conference on Prevention and Control of Oil Spills*, pp. 441–51. American Petroleum Institute, New York.

Zsolnay, A. (1971). Preliminary study of the dissolved hydrocarbons and hydrocarbons on particulate material in the Gotland Deep of the Baltic. *Kieler Meeresforschungen*, **21**, 55–80.

J.H.VANDERMEULEN & T.P.AHERN

Effect of petroleum hydrocarbons on algal physiology: review and progress report

The increase in bulk shipments of crude and refined oil is accompanied by the increasing risk of spills of large volumes of oil, particularly in coastal waters. The demonstrated toxicity of many components, some in extremely small concentrations, has spurred scientific concern and study of both short-term and long-term effects of exposure to petroleum hydrocarbons on plant and animal life. The bulk of this research has consisted of establishing toxicity levels and mortality rates. However, attention has recently turned to attempts to ascertain the basis of observed physiological effects, including metabolic target sites, both of whole oils and of the individual constituents of oil.

In this paper we briefly review the general physiological effects in unicellular algae of whole crude oils, distillate fractions and some individual constituents. We shall also consider in some detail the experimental problems associated with hydrocarbons, and the errors in interpretation resulting from these problems. Finally, we shall discuss more specifically the effects of one oil constituent, naphthalene, on the stability of the algal photosynthetic machinery and on ATP production.

Whole oils

The retardation of algal growth and the inhibition of algal photosynthesis by petroleum hydrocarbons has been amply demonstrated (Galtsoff, Prytherch, Smith & Koehring, 1935; Mironov & Lanskaya, 1969; Mommaerts-Billiet, 1973; Kauss & Hutchinson, 1974; Pulich, Winters & van Baalen, 1974; Soto, Hellebust, Hutchinson & Sawa, 1975*a*; Soto, Hellebust & Hutchinson, 1975*b*), both following direct mixing of oils into the medium, and with aqueous extracts. The patterns of retardation of cell growth, as measured by cell numbers, in algal cultures in these studies are, in general, similar. Both the initial lag phase in newly inoculated cultures and the subsequent rapid growth in the exponential phase are altered. The lag phase is lengthened, by several days in highly contaminated media, and the slope of the exponential phase is depressed. However, most cultures apparently return to normal growth ultimately if the medium is not too heavily contaminated.

Inhibition of DNA and RNA activity in three coastal macro-algae in the presence of relatively high concentrations of crude oil has recently been reported by Davavin, Mironov & Tsimbal (1975). A 24-hour exposure to emulsified oil–seawater mixtures (100 ppm–10000 ppm) resulted in decreased DNA in the red algae *Grateloupia dichotoma* and *Polysiphonia opaca*, and significantly reduced DNA and RNA specific activity in the green alga *Ulva lactuca*.

The distribution of photosynthetically fixed $^{14}CO_2$ following exposure to Kuwait crude oil has been examined by Brown (1972) in the marine lichen *Lichina pygmaea*. Treatment with fresh oil for 1 hour (unfortunately no concentration data are available) resulted in a slight but insignificant depression of ^{14}C fixed. Similarly no change was detectable in either the distribution of label into alcohol- and acid-soluble fractions, or in the amount of [^{14}C]glucose transferred from alga to fungus.

Susceptibility of algal species to oil varies, several species showing some resistance (*Chlamydomonas* sp., Nuzzi, 1973; certain species of *Ankistrodesmus*, Kauss & Hutchinson, 1974). Furthermore, stimulation of growth and of photosynthesis at lower oil concentration has also been noted both in pure algal cultures (Galtsoff *et al.*, 1935; Mironov & Lanskaya, 1969; Kauss & Hutchinson, 1974) and in natural phytoplankton stocks (Gordon & Prouse, 1973). Gordon & Prouse observed an enhancement of ^{14}C fixation in natural phytoplankton over a 24-hour incubation period in the presence of low concentrations ($<$ 50 ppb) of Venezuelan crude aqueous extract. However, with increased concentration (50 to 300 ppb) photosynthesis was depressed. Recently, Prouse, Gordon & Keizer (1975) found that both crude oil and No. 2 Fuel Oil, when added *after* the lag phase in concentrations similar to those reported in oil-contaminated waters (Levy, 1971; Michalik & Gordon, 1971), i.e. 19.3 to 786.5 ppb, did not significantly affect cell division of *Dunaliella* sp. and *Fragilaria* sp. over a period of 24 days.

This last study showed a significant technical improvement over most preceding work in two respects. Firstly, the actual concentrations of oil in the medium were measured and attempts were made to duplicate the low concentrations normally encountered in field situations; secondly, the authors demonstrated and measured the loss during the 14–18-day experiment of oil initially introduced. Evaporative loss was observed with both Kuwait crude oil and with No. 2 Fuel Oil and, in some cases, as much as 90 % of the amount introduced initially was lost over a 2-week period. The magnitude and rapidity of evaporative loss is indicated by an experiment using naphthalene (a bi-cyclic aromatic) alone. The concentration of naphthalene in a saturated aqueous

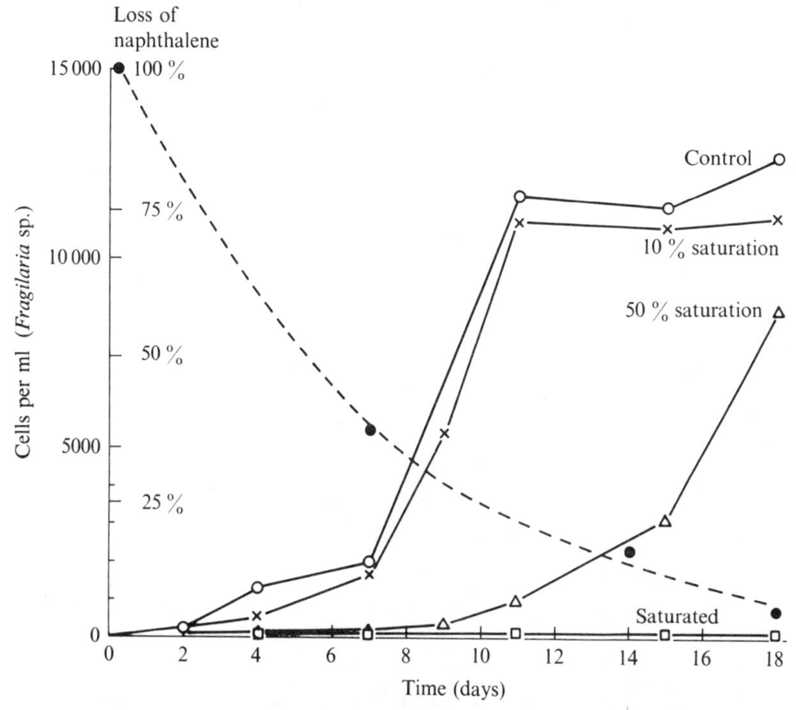

Fig. 1. Growth curves of *Fragilaria* sp. at different initial concentrations of naphthalene in f/2 medium (Guillard & Ryther, 1962). Algae were cultured axenically, at 10 °C, in 250-ml Erlenmeyer flasks positioned over banks of four fluorescent lamps (Sylvania F48T12–N–HO). Saturated naphthalene-medium solutions were prepared by stirring excess naphthalene in autoclaved f/2 medium for 3 to 5 days at 10 °C under sterile conditions. The saturated solution was filtered (Whatman cellulose filter, No. 4) under ultraviolet light, appropriate dilutions were made, and inoculated. Loss of naphthalene during this period was followed in a saturated control preparation (dashed line). Naphthalene concentrations were measured by fluorescence of hexane extracts in a Perkin-Elmer MPF-2A fluorescence spectrophotometer (excitation 310 nm, emission 336 nm).

solution in a cotton-plugged flask has been found to decrease rapidly with time; an initial loss of as much as 90 % occurring in the first 24 hours. Less than 5 % remains in solution after 10 days.

The composition of crude oils accommodated in seawater also changes with time. Prouse *et al.* (1975) report a considerable shift in the aromatic fraction and a simultaneous loss of n-alkanes below C_{20}.

Such changes in concentration or in the composition of oil or individual hydrocarbons during the course of an experiment will have significant effects on experimental studies, and indeed much of the growth recovery following

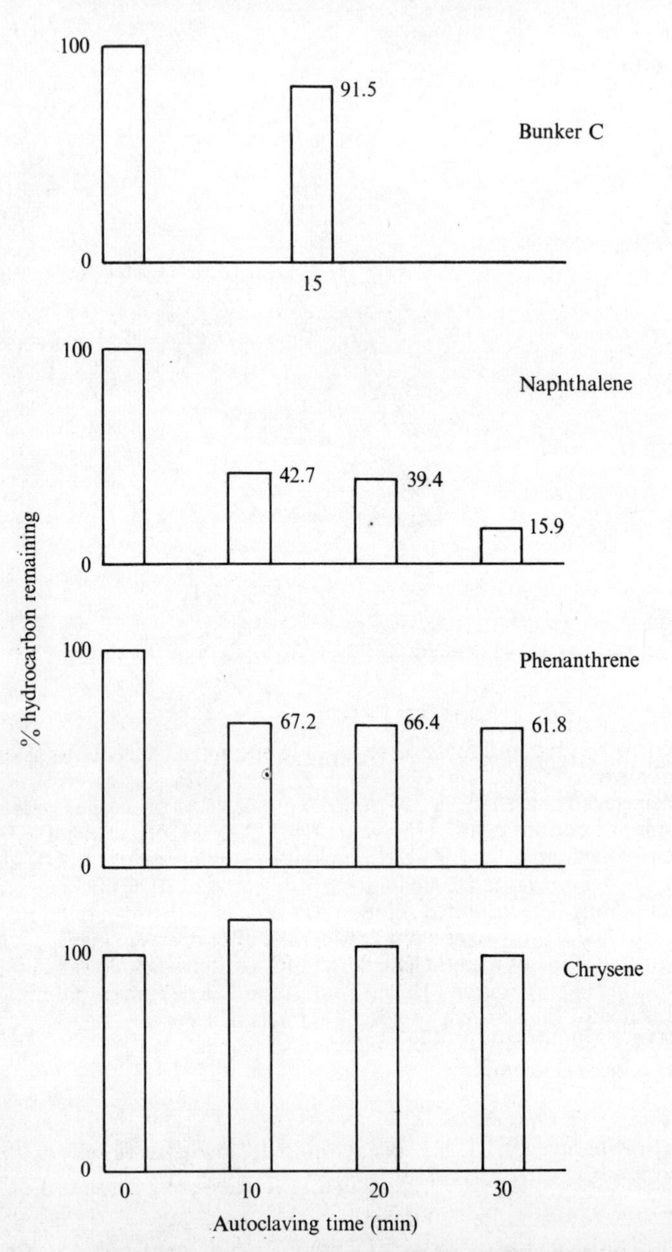

Fig. 2. The effect of autoclaving on concentrations of aromatic hydrocarbons and Bunker C oil dissolved in seawater (Barnstead Automatic Laboratory Sterilizer, 18 psi, 250 °C). Concentrations were determined by fluorescence spectroscopy of hexane extracts for Bunker C and naphthalene, and of cyclohexane extracts for phenanthrene and chrysene.

initial inhibition reported in the literature can be ascribed solely to decreasing concentrations and changing composition of hydrocarbons in the medium. Thus the recovery of cell division in a typical growth inhibition experiment of a culture of *Fragilaria* sp. in 50 % saturated naphthalene-seawater can be explained not only in terms of the cells overcoming the inhibiting effects of *c.* 750 ppb naphthalene, but also as a result of the concentrations decreasing to *c.* 200 ppb naphthalene (Fig. 1).

The main technical problem creating difficulties in interpretation of previous observations concerns the manner of preparation of oil extracts. Most of the research work on the toxic effects of oil has been done with pure algal species in axenic culture so as to avoid bacterial side-effects. Such axenic cultures require sterile media. However, since whole oils are complex populations of hydrocarbons, differing in concentration, in solubility and in volatility, sterilisation by boiling of an aqueous oil extract (Galtsoff *et al.*, 1935) is very suspect if it is hoped to maintain the composition of the oil. We ourselves have noted changes in aqueous extracts of Bunker C oil when autoclaved, and found both a decrease in the concentration of oil accommodated by seawater (Fig. 2). and a change in the aromatic composition as indicated by fluorescence intensity contour profiles (Fig. 3). An alternative sterilisation technique, by vacuum filtration through micropore membrane filters, has its own unique drawback. There is now evidence that oil in seawater does not exist solely in solution, but that the larger part (*c.* 90 %) accommodated exists in particulate form, the particles ranging from 1 to 30 μm in diameter (Gordon *et al.*, 1973). Thus sterilisation by vacuum filtration will effectively remove a large part of the oil present, thereby decreasing actual concentrations of oil even further. (This observation incidentally also changes our concept of actual solution concentrations encountered by algae in the field, even under heavily oiled conditions, these being probably much lower than hitherto expected.)

Autoclaving whole oils in bulk prior to equilibration with water does not appear to affect their composition. Prouse *et al.* (unpublished data) report no change in either aromatics, as indicated by synchronous excitation–emission spectra and fluorescence intensity contour profiles, or in relative concentrations of n-alkanes (by gas–liquid chromatography) in oil autoclaved in bulk.

Apart from difficulties arising from the failure to measure or report total or differential losses of hydrocarbons during experiments, further problems arise in respect of the manner of preparation of the oil extract. Indeed, in many studies neither the preparation technique nor the concentrations used are available (see Table 1). Where stated, oil–water mixtures range from emulsions to 10-minute shaking, to aqueous extracts prepared over several days, each method probably yielding its own final concentration and composition. Concentrations, where given, are generally expressed as a percentage of the

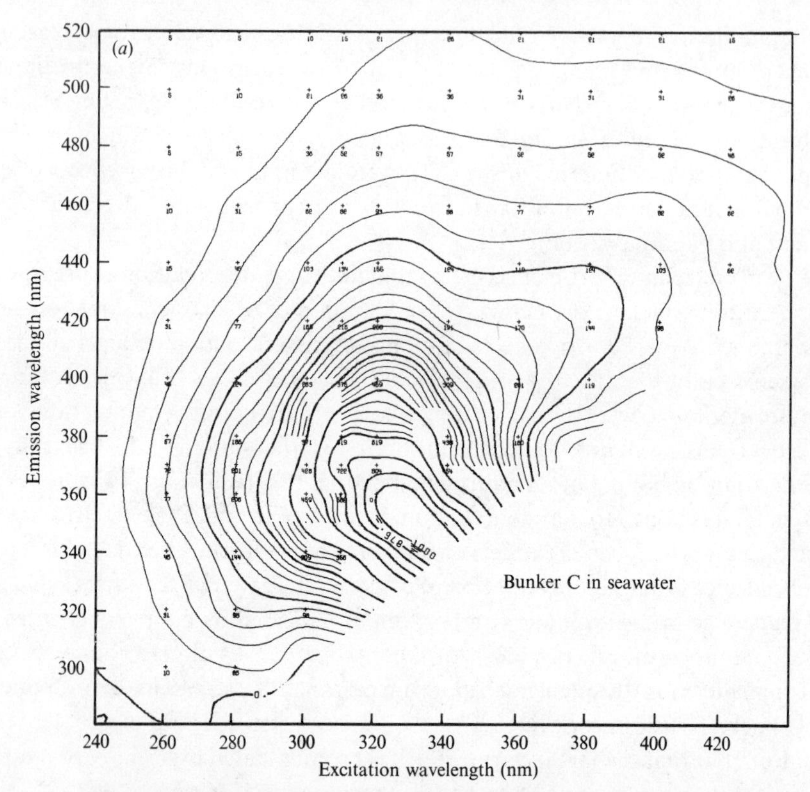

Fig. 3. Fluorescence intensity contour plots of seawater extracts of Bunker C oil, (*a*) before and (*b*) after autoclaving (15 min in a Barnstead Automatic Laboratory Sterilizer, 18 psi, 250 °C). Autoclaved solutions

concentration of the initial aqueous extract, but that initial concentration is rarely known. Also, the actual concentration in solution is probably very different from the amount of oil suspended in the medium. Thus a 1 ppm concentration of 1 ml of oil emulsified in 1 l of seawater cannot be compared with a 1 ppm aqueous extract, containing by actual measurement (fluorescence or gas–liquid chromatography) 1.0 ml of hydrocarbons in true solution. These details become especially relevant when attempting to relate or extrapolate from observations obtained in the laboratory to problems in the field.

Distillate fractions

As emphasised above, a major problem with crude or refined oil studies is the question of the hydrocarbon composition of the oils and specific effects of

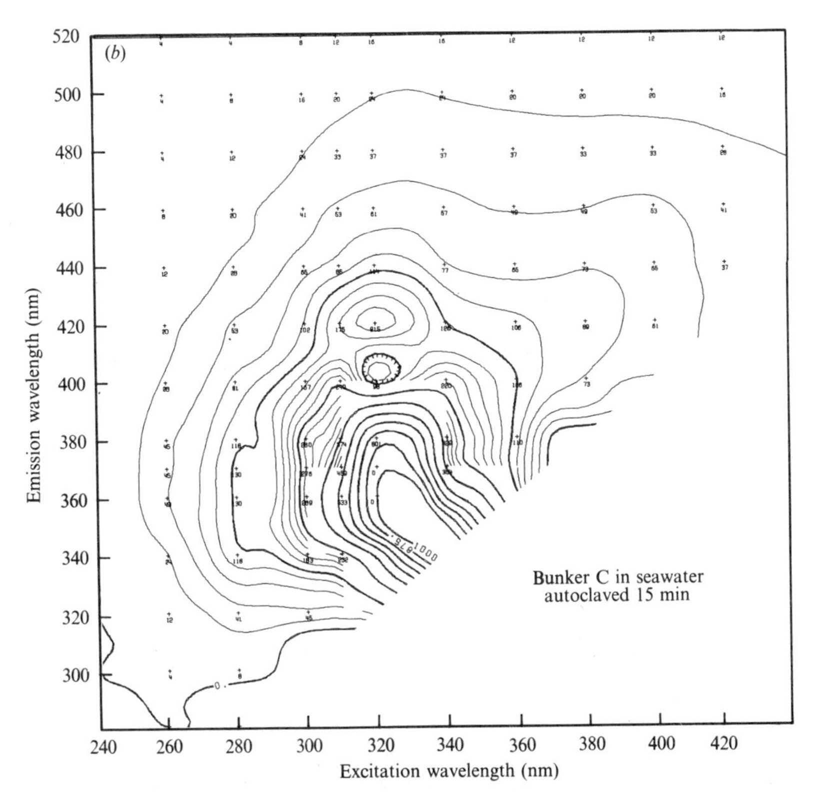

were allowed to cool to room temperature before extraction with redistilled hexane. Two-dimensional plots were prepared by the method of Hargrave & Phillips (1975).

certain fractions. For this reason workers have been turning either to standard oils of known composition (Prouse & Gordon, 1974; Pulich *et al.*, 1974) (available from the American Petroleum Institute reference collection, Dr J. Anderson, Texas A & M University), or to specific distillate fractions of oils (Pulich *et al.*, 1974) and to single known oil constituents such as naphthalene (Kauss *et al.*, 1973; Stoll & Guillard, 1974; Soto *et al.*, 1974, 1975*a,b*).

Studies with certain distillate fractions of No. 2 Fuel Oil have demonstrated interesting specific susceptibilities of certain algae (Pulich *et al.*, 1974). Thus, growth in the green alga *Chlorella autotrophica* was severely inhibited by lower-boiling-point fractions, although a general toxicity was exerted by all fractions. In contrast, the blue-green alga *Agmenellum quadruplicatum* and the diatom *Thalassiosira pseudonana* were found to be most susceptible to the high-boiling fractions. However, aqueous extracts of Kuwait and Southern Louisiana crude oils did not appear to affect algal growth, but specific fractions of Southern Louisiana crude, tested separately, showed toxicity

Table 1. *Summary of studies concerning effects of petroleum hydrocarbons on algal growth, photosynthesis and metabolism, in fresh and marine waters*

Compound	Preparation of aqueous extract	Hydrocarbon concentration in medium	Algal species	Exposure time	Effect (growth, O_2, $^{14}CO_2$ fixation)	Reference
Oils						
Pelto crude Mineral oil Cod-liver oil	Stirring, aqueous	[1]	*Nitzschia* sp.	5–21 d	Inhibition of growth	[5a]
Whole oils [2]	Not available	[1]	*Chaetoceros* sp. *Ditylum* sp. *Licmophora* sp. *Coscinodiscus* sp. *Glenodinium* sp. *Gymnodinium* sp. *Prorocentrum* sp. *Peridinium* sp. *Platymonas* sp.	5 d	Inhibition of growth	[5b]
Libyan crude	Not available	[1]	*Platymonas* sp.	17 d	Diminished growth	[5c]
Venezuela crude, No. 2 Fuel Oil, No. 6 Fuel Oil	Shaking 10 min; aqueous phase used directly	0–300 ppb[3]	Natural phytoplankton population	6–7 h	Carbon-fixation stimulated at low concentrations, depressed at high concentrations. No. 2 Fuel Oil most toxic	[5d]
Venezuela crude, Kuwait crude[4], No. 2 Fuel Oil[4]	Oil and seawater autoclaved separately; shaken; aqueous extract used directly	2–787 ppb[3]	*Dunaliella* sp. *Fragilaria* sp.	14–24 d	No effect on growth	[5e]
Norman Wells crude	Stirring 12-h; aqueous phase used directly	[1]	*Ankistrodesmus* spp.	10 d	Range of susceptibility	[5f]
Kuwait crude[4], Southern Louisiana crude, No. 2 Fuel Oil[4]	Both whole oils and distillate fractions; stirred; aqueous extract filtered (0.45 μm)	40–400 ppb[6]	*Thalassiosira* sp. *Agmenellum* sp. *Chlorella* sp.	7 d	No. 2 toxic to growth and O_2 output in *Thalassiosira* sp., lesser degree in others. Varied response to intact and aqueous crude extracts	[5g]

Romaskinskaya oil	Emulsification in seawater	[1]	*Ulva* sp. *Grateloupia* sp. *Polysiphonia* sp.	24 h	Modification of polymerisation of DNA, inhibition of biosynthesis of DNA and RNA	[5h]
No. 2 Fuel Oil, No. 6 Fuel Oil, outboard motor oil	Shaking 1 % oil-medium 1 h; phase separation and micro-filtration	[1]	*Phaeodactylum* sp. *Skeletonema* sp. *Chlorella* sp. *Chlamydomonas* sp.	10–12 d	No. 2 Fuel Oil inhibited growth in *Phaeodactylum*, *Chlorella*, *Skeletonema*; *Chlamydomonas* was resistant. No. 6 and motor oil little effect	[5i]
Aromatics						
7 Western oils, outboard motor oil, benzene, toluene, o-xylene, naphthalene	Stirring 12 h; aqueous phase used directly	[1]	*Chlorella* sp.	10 d	Inhibition of growth	[5j]
Naphthalene	Stirring 12 h; aqueous phase used directly	[1]	*Chlamydomonas* sp.	16 d	Depression of carbon fixation	[5k]
Naphthalene and British Petroleum mixed sour oil	Not available	[1]	*Chlamydomonas* sp.	6–18 d	Naphthalene inhibited carbon fixation; sour oil less severe	[5l]
Naphthalene	Dissolution at 80 °C; used directly	[1]	*Chlamydomonas* sp.	18 d	Inhibition of growth	[5m]
Carcinogenic and non-carcinogenic polycyclics	0.1 % solutions in methyl cellosolve	3 ppb–290 ppm	*Antithamnion* sp. *Spermothamnion* sp. *Callithamnion* sp.	0.5–96 h	Growth stimulation, with cell size reduction	[5n]

[1] Concentrations provided in literature expressed as fraction of aqueous extract.
[2] Paper in Russian. No further information available.
[3] Concentration in culture medium determined by fluorescence spectroscopy.
[4] Obtained from American Petroleum Institute reference collection, available from Dr J. Anderson, Texas A&M University.
[5] (*a*) Galtsoff *et al.* (1935); (*b*) Mironov & Lanskaya (1969); (*c*) Mommaerts-Billiet (1973); (*d*) Gordon & Prouse (1973); (*e*) Prouse & Gordon (1974); (*f*) Kauss & Hutchinson (1974); (*g*) Pulich *et al.* (1974); (*h*) Davavin *et al.* (1975); (*i*) Nuzzi (1973); (*j*) Kauss *et al.* (1973); (*k*) Kauss *et al.* (1973); (*l*) Soto *et al.* (1974); (*m*) Soto *et al.* (1975*a*); (*n*) Boney & Corner (1962); Boney (1974).
[6] Workers' estimate.

both to the diatom and the green alga. Curiously, both Kuwait and Southern Louisiana crude, when added directly to the cultures rather than as aqueous extracts, were toxic to varying degrees. We thus observe a range of responses, with species-specific toxicities residing in certain fractions and toxicity of crudes varying according to whether the crudes are introduced wholly or as aqueous extracts.

Aromatic hydrocarbons

Recently interest has centred on the lower-boiling-point volatile aromatic hydrocarbons because: (*a*) much of the toxicity of crude oils appears to be associated with this group of compounds (Nelson-Smith, 1968; Baker, 1970; Boyland & Tripp, 1971; Soto *et al.*, 1975*a,b*); (*b*) they constitute an important part of aqueous oil extracts (Boylan & Tripp, 1971); and (*c*) they show considerable solubility in both fresh and salt water (McAuliffe, 1966; Mackay & Wolkoff, 1973).

The carcinogenicity of various poly-cyclic aromatic hydrocarbons has been examined in red algae (Boney & Corner, 1962; Boney, 1974). It was found in experiments with filamentous sporelings that apically mediated growth was significantly enhanced by contact with these compounds, even after brief exposure times (< 1 hour). However, this was accompanied by reduction in cell size. This growth stimulation has been noted with both known carcinogenic as well as with non-carcinogenic aromatics, e.g. chrysene and 9-methyl-anthracene. It is tempting to speculate whether the reported stimulation of growth in phytoplankton by low oil concentrations, generally thought to be due to utilisation of the hydrocarbon as a metabolic substrate, may in fact be partly as a result of slight carcinogenic stimulatory activity.

Benzene and its related compounds, naphthalene and methylated derivatives, and phenanthrene, all show varying degrees of inhibition of growth both in the green alga *Chlorella vulgaris* and in the blue-green *Agmenellum* sp. (Kauss *et al.*, 1973; Pulich *et al.*, 1974). They can be ranked tentatively as to toxicity as follows: benzene, toluene, xylene, phenanthrene, naphthalene, cresol, trimethyl benzene, methyl naphthalene, and dimethyl naphthalene.

Naphthalene, because it occupies a central position in the toxicity ranks, has been used in recent studies. It appears to be readily taken up from solution, even at high concentrations, and is inhibiting to both cell growth and photosynthesis (Kauss *et al.*, 1973; Soto *et al.*, 1975*a,b*). Both cell division and carbon fixation are regained, however, when cells are resuspended in naphthalene-free medium or after aeration of the culture flasks sufficient to decrease naphthalene in solution. It is still uncertain whether cells have the ability to rid themselves of naphthalene, either by excretion or by metabolism. Disappearance of [^{14}C]naphthalene from cells appears to be related to the onset of cell division. Thus labelled cells, when resuspended in a naphthalene-free medium, began to lose label immediately and regained cell division within 48 hours. However, naphthalene-labelled cells, when left in the naphthalene-dosed medium, did not begin to lose label until cell division began, some 4 days after aeration of the culture flasks. These observations have been interpreted to mean that, at least in *Chlamydomonas* sp., naphthalene persists in the population, being diluted only by cell division. Whether algae possess hydrocarbon-metabolising systems similar to the cytochrome-P_{450} system in animals is still a moot point. Soto *et al.* (1975*b*) were unable to detect non-volatile labelled components in recovery medium after resuspension of [^{14}C]naphthalene-labelled *Chlamydomonas* cells. They interpreted this to mean that *Chlamydomonas* is incapable of metabolising simple aromatics.

General changes in protein, carbohydrate, lipid and photopigments have been reported for *Chlamydomonas* sp. grown for 7 days in 50% naphthalene-saturated medium before resuspension in naphthalene-free recovery medium (Soto *et al.*, 1974). Cellular total protein decreased during the initial 7-day incubation period, but was restored within 24 hours in recovery medium. In contrast, both lipid and carbohydrate reportedly increased more than twofold in the presence of naphthalene, but returned to normal levels in the recovery medium. Total pigment concentration did not apparently change during the initial incubation with naphthalene, but increased appreciably afterwards in the recovery medium.

The use of naphthalene as a model aromatic hydrocarbon brings with it its inherent experimental and interpretation problems. Although soluble in water (Table 2) saturation is achieved only after prolonged stirring. Furthermore, little information on solubility and solubilising behaviour is available, particularly with respect to seawater or seawater medium. Also, it is highly volatile and, as has been pointed out earlier, disappears readily from solution.

There are two means of introducing naphthalene into an algal culture. The first, as used by Kauss *et al.* (1973), Soto *et al.* (1974, 1975*a*,*b*) and by ourselves (viz. Fig. 1 of this paper), is that of growing or suspending algal cells in medium already presaturated with naphthalene by prolonged stirring. Intermediate concentrations of naphthalene are then obtained by appropriate dilutions of saturated medium with naphthalene-free medium. The main disadvantage of this method lies in the time-consuming process of saturating the culture medium, and later in centrifuging and resuspending algal cells from stock culture. The alternative method consists of injecting naphthalene dissolved in a suitable organic solvent, such as acetone or ethyl alcohol. The advantage of this method lies in the ease of preparing a large number of

Table 2. *Solubility of naphthalene in distilled water and in filtered seawater (0.45 μm HAWP, Millipore)*

Solvent	Temperature (°C)	Concentration of naphthalene (mg l^{-1})	Reference
Distilled water	25	33	Mackay & Wolkoff (1973)
Distilled water	25	34.4	Bohon & Claussen (1951)
Seawater	20	23.1	Vandermeulen & Ahern (unpublished data)
Seawater	10	12.7	Vandermeulen & Ahern (unpublished data)

Solubility in seawater was determined by fluorescence of hexane extracts (excitation 310 nm, emission 325 and 336 nm).

Table 3. *Solubility of naphthalene in filtered seawater at* 10°C *(0.45 μm HAWP, Millipore) when added pre-dissolved in acetone*

Concentration added dissolved in acetone* (mg l^{-1})	Concentration measured in solution† (mg l^{-1})	% dissolved
1	0.84	85
5	3.90	78
10	5.36	54
15	5.92	40
20	7.10	36

* Naphthalene-acetone was added so as to yield a final volume of 1 ml acetone per litre of filtered seawater.
† Average of three replicates. The solution was allowed to stand for 15 min before filtration through Whatman No. 4 cellulose filter paper to remove naphthalene crystals in suspension. Solutions were extracted into redistilled hexane and naphthalene concentration determined by fluorescence (excitation 310 nm, emission 336 nm).

experimental replicates, and in the ease of experimental manipulation. The disadvantage of course is in possible side-effects or synergistic effects of the solvent. Therefore the amount of solvent used as a carrier must be kept to a minimum. A second problem is the solubility behaviour of naphthalene in seawater in the presence of an organic solvent such as acetone (viz. Table 3). The actual amount of naphthalene measured in solution is less than and varies with the concentration of naphthalene in the acetone injected.

We have examined naphthalene-mediated inhibition of carbon fixation and

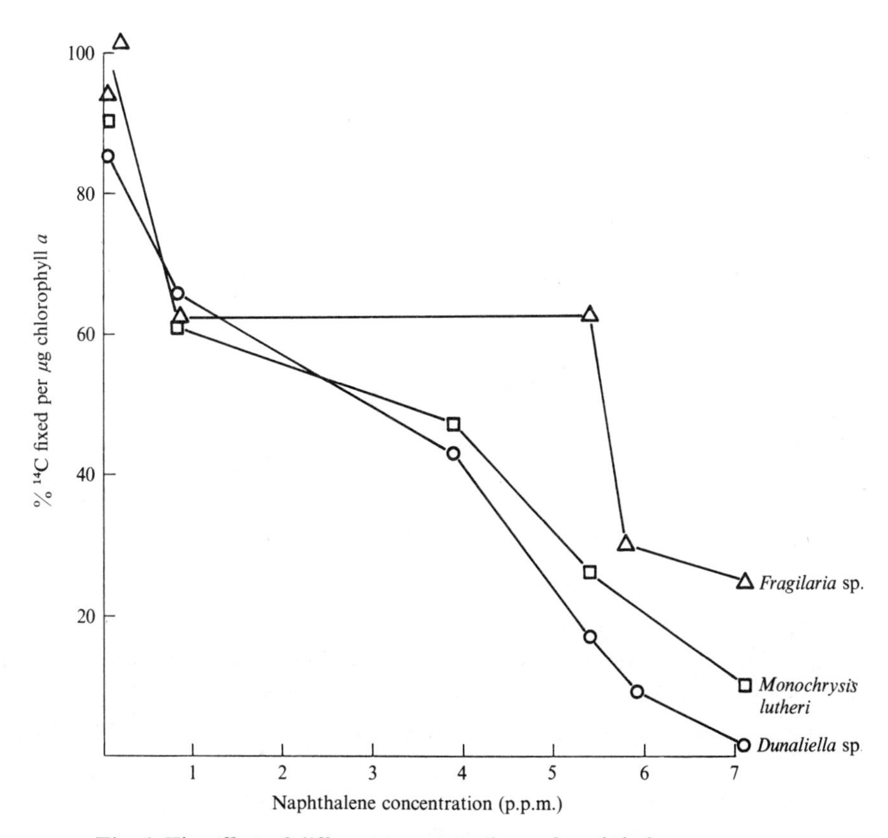

Fig. 4. The effect of different concentrations of naphthalene on photosynthesis in three marine unicellular algae. Naphthalene, dissolved in acetone (final concentration of acetone, 1 μl ml^{-1} culture) was added to 1-ml volumes of algal cultures and photosynthesis was monitored by $H^{14}CO_3$ fixation (Steeman-Nielsen, 1952). No significant differences in photosynthesis were found in control cultures containing acetone only. Experimental cultures were pre-incubated with naphthalene for 20 min before addition of isotope (New England Nuclear, 1 μCi ml^{-1} distilled water, $H^{14}CO_3$). Algae were harvested by vacuum filtration onto membrane filters (0.45 μm, HAWP, Millipore). Filters were dried gently over low heat (38 °C, 15 min), fumed over concentrated hydrochloric acid (15–20 min), and counted in Permafluor-toluene (Packard). Radioactivity was determined in a Nuclear Chicago Isocap 300 liquid scintillation counter. Chlorophyll *a* was determined by the method of Yentsch & Menzel (1963) using a Turner Model 110 fluorometer.

the stability of the photosynthetic machinery in some representative marine unicellular algae. Depression of carbon-fixation appears to be dose-dependent, and the pattern of inhibition is similar in the three algae studied, the diatom *Fragilaria* sp., *Monochrysis lutheri* and *Dunaliella* sp. (Fig. 4). There was an initial large decrease in ^{14}C fixation. Between 0 and 1 ppm approximately 70 %

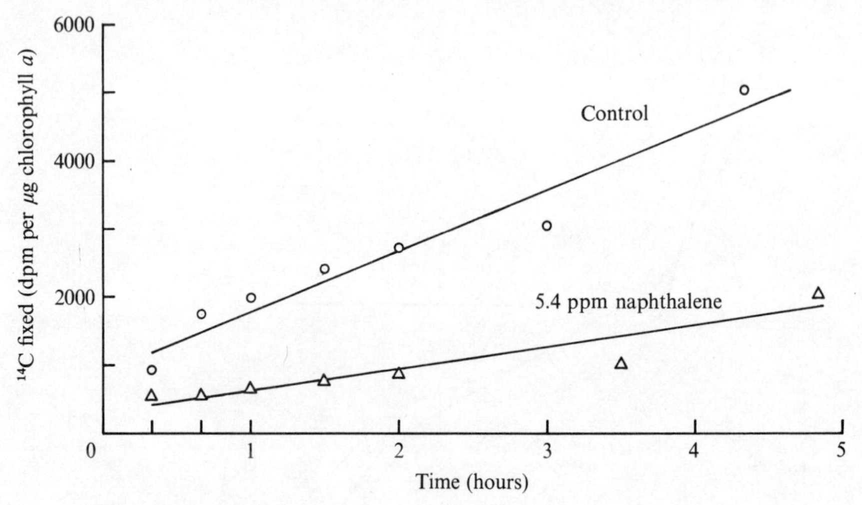

Fig. 5. The effect of 5.4 ppm naphthalene on photosynthesis over time in *Monochrysis lutheri*.

of photosynthesis was suppressed and at 20 ppm over 80 % inhibition of photosynthesis was observed. A study of the time course of inhibition at ~ 5 ppm (Fig. 5) shows (*a*) that the amount of inhibition does not appear to change with time, remaining approximately 70 % throughout, but (*b*) that despite this inhibition, carbon fixed continues to rise with time. When cells are resuspended in naphthalene-free medium after 4 hours' pre-incubation in ~ 5 ppm naphthalene, carbon fixation immediately increases, reaching normal levels within 5 hours (Fig. 6). Over this same time period we found no changes in either chlorophyll *a* content, or of its breakdown product phaeophytin (Table 4). However, more significantly ATP content was decreased both in the light and in the dark during this period in cells incubated with 5 ppm naphthalene (Table 5).

These various observations suggest:

(*a*) that marine unicellular algae are highly sensitive to small changes in trace amounts of naphthalene, and possibly other aromatic hydrocarbons;

(*b*) that the mechanism of inhibition of carbon fixation appears to be bimodal, with differential inhibition at low and high concentrations;

(*c*) that in short-term exposures (less than 12 hours) the inhibitory effect is reversible, implying reversible binding of the naphthalene molecule;

(*d*) that the photosynthetic machinery itself remains functional at high concentrations of naphthalene over this time period;

(*e*) that the observed depression of carbon fixation may well be a secondary effect of inhibition elsewhere in the cellular machinery, such as blocking

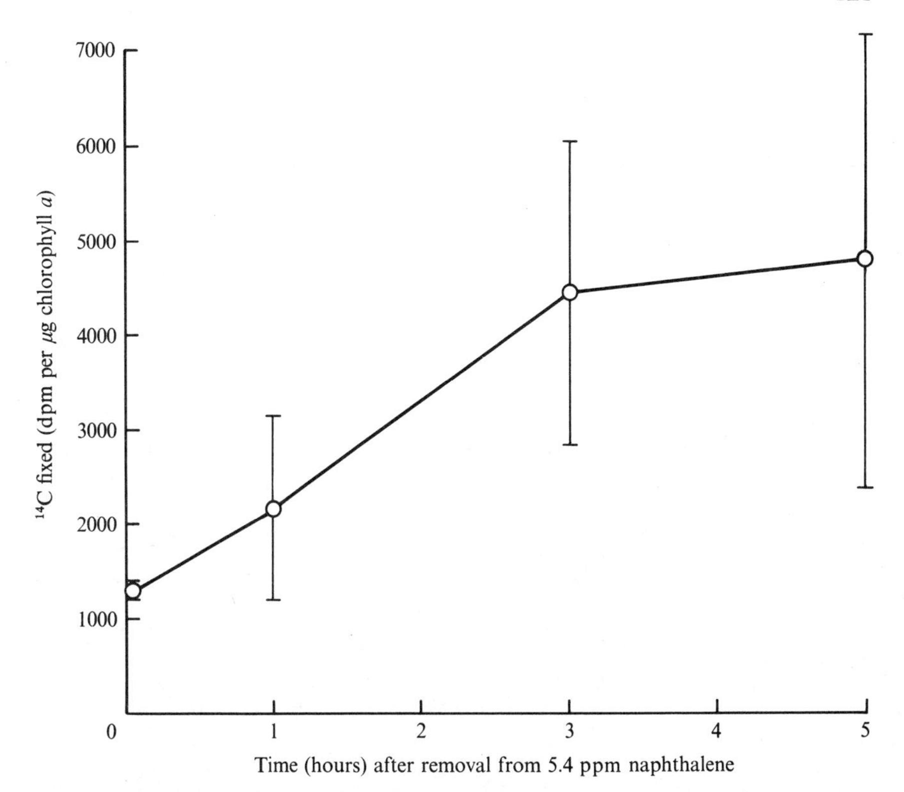

Fig. 6. Reversibility of inhibition of photosynthesis in *Monochrysis lutheri*. Cultures were pre-incubated with 5.4 ppm naphthalene for 4 hours, centrifuged gently, and resuspended in naphthalene-free f/2 medium before incubation with [^{14}C]bicarbonate.

of oxidative phosphorylation. The fact that ATP content is reduced in the dark, in the absence of photosynthesis, suggests that this may well be the case.

Of how much relevance are these observations to the ecologist, or to the field situation, obtained as they were largely with axenic cultures? They probably have little direct relationship to open ocean oil spill conditions since, aside from the obvious dilution effect, the minimum concentration of naphthalene used experimentally to obtain a quantifiable effect (1.0 ppm) is higher than those reported in aqueous extracts of oils (0.153 ppm in seawater extracts of kerosene and 0.0323 ppm in seawater extracts of Kuwait crude oil; Boylan & Tripp, 1971). The same argument holds for most of the studies using aqueous extracts of whole oils. Although in general concentrations of oil in these aqueous extracts were not available, estimates place them well above those normally encountered in oil-contaminated waters. However,

Table 4. *Chlorophyll* a *(C) and phaeophytin (P) concentrations in* Monochrysis lutheri *exposed to naphthalene for up to 7 hours*

Naphthalene concentration in solution*	Pigment ($\mu g\ ml^{-1}$ culture)	Exposure time (hours) 1	2	4	7
f/2 control	C	0.19±0.00			
	P	0.04±0.00			
	P/C	0.23			
Acetone control	C	0.21±0.00	0.19±0.01	0.20±0.01	0.19±0.01
	P	0.05±0.00	0.05±0.01	0.04±0.01	0.05±0.01
	P/C	0.23	0.26	0.21	0.26
3.9 ppm	C	0.21±0.00	0.20±0.01	0.21±0.01	0.20±0.01
	P	0.04±0.01	0.03±0.01	0.03±0.02	0.03±0.01
	P/C	0.20	0.13	0.12	0.16
5.36 ppm	C	0.19±0.02	0.20±0.01	0.19±0.01	0.19±0.02
	P	0.05±0.02	0.03±0.01	0.03±0.01	0.06±0.03
	P/C	0.26	0.17	0.17	0.29
7.10 ppm	C	0.16±0.09	0.19±0.01	0.20±0.01	0.19±0.01
	P	0.03±0.04	0.04±0.01	0.04±0.01	0.06±0.02
	P/C	0.21	0.21	0.19	0.29

Values are the mean ± 1 S.D.
* Appropriate stock solutions of naphthalene dissolved in acetone were added so as to yield the final concentrations shown. Concentrations determined by fluorescence (viz. Table 3).

Table 5. *Concentration of ATP in* Monochrysis lutheri *in f/2 medium with* 5.4 *ppm naphthalene added*

	ATP ($m\mu g\ \mu g^{-1}$ chlorophyll *a*)			
Experimental conditions	No. of tests	Light	No. of tests	Dark
Control	4	40.24±4.06	4	38.40±2.65
Naphthalene	4	26.56±4.89	4	20.52±7.11

Values are the mean ±1 S.D.
ATP measured by the method of Holm-Hansen & Booth (1966).

these various experimental observations become important when dealing with oil spill situations with lowered evaporative loss and at lower temperatures, both of which conditions decrease volatilising of the toxic naphthalenes and other aromatics, thereby keeping them in solution. They are also relevant to spills in restricted areas with little or no dilution, as in bays, inlets and inland water systems. A continuous, long-term source of petroleum hydrocarbons not recognised until recently is the essentially unaltered oil from spills stranded on low-energy beaches (Rashid, 1974) and imbibed into the sediments (Blumer, Ehrhardt & Jones, 1973). All these conditions lead to increased exposure of algae to hydrocarbons. Another potentially damaging situation is that of a spill in an already stressed community or environment, as suggested by the preliminary study of Stoll & Guillard (1974) who observed that batch-cultured marine diatoms deficient in phosphorus were more sensitive to naphthalene (i.e. showed a greater reduction of growth rate) than were control populations with higher cell phosphorus.

The experimental evidence therefore suggests that single hydrocarbons can affect the metabolic processes of algae. However, the real threat of hydrocarbons to living systems probably lies not in the toxicity of a single type, but in the synergistic effects of several acting together.

We gratefully acknowledge the advice on algal cultures by Dr Subba Rao Durvasula who supplied the inocula, and the technical assistance of Mr J. Hanrahan. For assistance in ATP assays we thank Dr W. H. Sutcliffe, Jr and Mrs A. Orr. We thank Dr D. C. Gordon, Dr B. T. Hargrave and Dr Durvasula for critically reading the manuscript. We are grateful to the Society of Experimental Biology, UK, for inviting us to contribute to this volume.

References

Baker, J. M. (1970). Oil pollution in salt-marsh communities. *Marine Pollution Bulletin*, **1**, 27–8.

Blumer, M., Ehrhardt, M. & Jones, J. H. (1973). The environmental fate of stranded crude oil. *Deep-Sea Research*, **20**, 239–59.

Bohon, R. L. & Claussen, W. F. (1951). The solubility of aromatic hydrocarbons in water. *Journal of the American Chemical Society*, **73**, 1571–8.

Boney, A. D. & Corner, E. D. S. (1962). On the effects of some carcinogenic hydrocarbons on the growth of sporelings of marine red algae. *Journal of the Marine Biological Association of the United Kingdom*, **42**, 579–85.

Boney, A. D. (1974). Aromatic hydrocarbons and the growth of marine algae. *Marine Pollution Bulletin*, **5**, 185–6.

Boylan, D. B. & Tripp, B. W. (1971). Determination of hydrocarbons in seawater extracts of crude oil and crude oil fractions. *Nature, London*, **230**, 44–7.

Brown, D. H. (1972). The effect of Kuwait crude oil and a solvent emulsifier on the metabolism of the marine lichen *Lichina pygmaea*. *Marine Biology*, **12**, 309–15.

Davavin, I. A., Mironov, O. G. & Tsimbal, O. G. (1975). Influence of oil on nucleic acids of algae. *Marine Pollution Bulletin*, **6**, 13–15.

Galtsoff, P. S., Prytherch, H. F., Smith, R. O. & Koehring, V. (1935). Effects of crude oil pollution on oysters in Louisiana waters. *Bulletin of the Bureau of Fisheries, Washington*, **18**, 143–210.

Gordon, D. C. Jr, Keizer, P. D. & Prouse, N. J. (1973). Laboratory studies of the accommodation of some crude and residual fuel oils in sea water. *Journal of the Fisheries Research Board of Canada*, **30**, 1611–18.

Gordon, D. C. & Prouse, N. J. (1973). The effects of three oils on marine phytoplankton photosynthesis. *Marine Biology*, **22**, 329–33.

Guillard, R. R. L. & Ryther, J. H. (1962). Studies of marine planktonic diatoms. 1. *Cyclotella nana* Hustedt and *Detonula confervacea* (Cleve) Gran. *Canadian Journal of Microbiology*, **8**, 229–39.

Hargrave, B. T. & Phillips, G. A. (1975). Estimates of oil in aquatic sediments by fluorescence spectroscopy. *Environmental Pollution*, **8**, 193–215.

Holm-Hansen, O. & Booth, Ch. R. (1966). The measurement of adenosine triphosphate in the ocean and its ecological significance. *Limnology and Oceanography*, **11**, 510–19.

Kauss, P. B. & Hutchinson, T. C. (1974). Studies on the susceptibility of *Ankistrodesmus* species to crude oil components. In *Proceedings of the 19th congress of the International Association of Theoretical and Applied Limnology, Winnipeg, Canada*.

Kauss, P. B., Hutchinson, T. C., Soto, C., Hellebust, J. & Griffiths, M. (1973). The toxicity of crude oil and its components to freshwater algae. In *Conference on prevention and control of oil spills, 1973, Washington, DC*, pp. 703–14.

Levy, E. M. (1971). The presence of petroleum residues off the east coast of Nova Scotia, in the Gulf of St Lawrence, and the St Lawrence River. *Water Research*, **5**, 723–33.

McAuliffe, C. (1966). Solubility in water of paraffin, cycloparaffin, olefin, acetylene, cycloolefin, and aromatic hydrocarbons. *Journal of Physical Chemistry*, **70**, 1267–75.

Mackay, D. & Wolkoff, A. W. (1973). Rate of evaporation of low-solubility contaminants from water bodies to atmosphere. *Environmental Science and Technology*, **7**, 611–14.

Michalik, P. A. & Gordon, D. C. Jr (1971). *Concentration and distribution of oil pollutants in Halifax Harbour 10 June to 20 August, 1971*. Technical Report No. 284, Fisheries Research Board of Canada. 25 pp.

Mironov, O. G. & Lanskaya, L. A. (1969). Growth of marine microscopic algae in seawater contaminated with hydrocarbons. *Biologiva Morya*, **17**, 31–8. (In Russian.)

Mommaerts-Billiet, F. (1973). Growth and toxicity tests on the marine nanoplanktonic alga *Platymonas tetrathele* G. S. West in the presence of crude oil and emulsifiers. *Environmental Pollution*, **4**, 261–82.

Nelson-Smith, A. (1968). Biological consequences of oil pollution and shore cleansing. In *The biological effects of oil pollution on littoral communities*, pp. 73–80. Field Studies Council, London.

Nuzzi, R. (1973). Effects of water soluble extracts of oil on phytoplankton. In *Conference on prevention and control of oil spills, 1973, Washington, DC*, pp. 809–13.

Prouse, N. J. & Gordon, D. C. Jr (1974). *The effects of three oils on the growth of the dinoflagellate* Dunaliella tertiolucta *and the diatom* Fragilaria *sp. in axenic batch cultures*. International Council for the Exploration of the Sea, C.M. 1974/2:41. 6 pp.

Prouse, N. J., Gordon, D. C. Jr & Keizer, P. D. (1975). The effects of oil-contaminated seawater on the exponential growth of unialgal batch cultures of marine phytoplankton. *Journal of the Fisheries Research Board of Canada*, submitted.

Pulich, W. M. Jr, Winters, K. & Van Baalen, C. (1974). The effects of a No. 2 fuel oil and two crude oils on the growth and photosynthesis of microalgae. *Marine Biology*, **28**, 87–94.

Rashid, M. A. (1974). Degradation of Bunker C oil under different coastal environments of Chedabucto Bay, Nova Scotia. *Estuarine and Coastal Marine Science*, **2**, 137–44.

Soto, C., Hellebust, J. A. & Hutchinson, T. C. (1974). The effects of aqueous extracts of crude oil and naphthalene on the physiology and morphology of a freshwater green alga. In *Proceedings of the 19th congress of the International Association of Theoretical and Applied Limnology, Winnipeg, Canada.*

Soto, C., Hellebust, J. A. & Hutchinson, T. C. (1975*b*). Effect of naphthalene and aqueous crude oil extracts on the green flagellate *Chlamydomonas angulosa*. II. Photosynthesis and the uptake and release of naphthalene. *Canadian Journal of Botany*, **53**, 118–26.

Soto, C., Hellebust, J. A., Hutchinson, T. C. & Sawa, T. (1975*a*). Effect of naphthalene and aqueous crude oil extracts on the green flagellate *Chlamydomonas angulosa*. I. Growth. *Canadian Journal of Botany*, **53**, 109–17.

Steemann-Nielsen, E. (1952). The use of radioactive carbon for measuring organic production in the sea. *Journal du Conseil, Conseil Permanent pour l'Exploration de la Mer*, **18**, 117–40.

Stoll, D. R. & Guillard, R. R. L. (1974). Synergistic effect of naphthalene, toxicity and phosphate deficiency in a marine diatom. *Abstracts, 37th Annual Meeting, American Society of Limnology and Oceanography.*

Yentsch, C. S. & Menzel, D. W. (1963). A method for the determination of phytoplankton chlorophyll and phaeophytin by fluorescence. *Deep-Sea Research*, **10**, 221–31.

1

R.F.ADDISON

Organochlorine compounds in aq organisms: their distribution, transport and physiological significance

Introduction

This paper is about the distribution and metabolism of three groups of chlorinated hydrocarbons in aquatic organisms, and about some of their effects. These materials include (*a*) the DDT* group of insecticides, (*b*) the polychlorinated biphenyls (PCBs) and (*c*) the cyclodiene insecticides. I have restricted my discussion to these materials since most information is available about them. Although other chlorinated organic compounds such as chlorinated paraffins or the chlorophenoxyacetic acids may be as important industrially or agriculturally as those selected, much less information about their behaviour exists, and any discussion of their significance to the aquatic (and particularly the marine) environment would be even more speculative than this review will be. I have deliberately omitted any discussion of the significance of chlorinated hydrocarbon compounds to birds, on the grounds that this subject is too large to summarise adequately in a review such as this.

The chlorinated hydrocarbons (OC) are 'heavy' chemicals, i.e. they have been produced on a large scale for several decades (e.g. more than 5×10^7 kg each in 1965 in the US alone: Anon., 1969; Hutzinger, Safe & Zitko, 1974). The PCBs were introduced first in the late 1920s and have been used in a variety of applications where their particular properties of chemical unreactivity and availability in a range of physical states are desirable; these applications include uses as heat transfer fluids, condenser dielectrics and hydraulic fluids. They are prepared industrially by the chlorination of biphenyl, a reaction which yields a wide range of compounds differing in the number and position of chlorine substituents on the biphenyl molecule. PCBs are produced as distillation fractions of the refined reaction products and each fraction contains a number of compounds. There are special

* The following abbreviations are used: *p,p'*-DDT, 2,2-bis-(*p*-chlorophenyl)-1,1,1-trichloroethane; *p,p'*-DDD, 2,2-bis-(*p*-chlorophenyl)-1,1-dichloroethane; *p,p'*-DDMS, 2,2-bis-(*p*-chlorophenyl)-1-chloroethane; *p,p'*-DDE, 2,2-bis-(*p*-chlorophenyl)-2,2-dichloroethylene; *p,p'*-DDMU, 2,2-bis-(*p*-chlorophenyl)-1-chloroethylene; *p,p'*-DDA, 2,2-bis-(*p*-chlorophenyl) acetic acid; *p,p'*-DBP, 4,4'-dichlorobenzophenone.

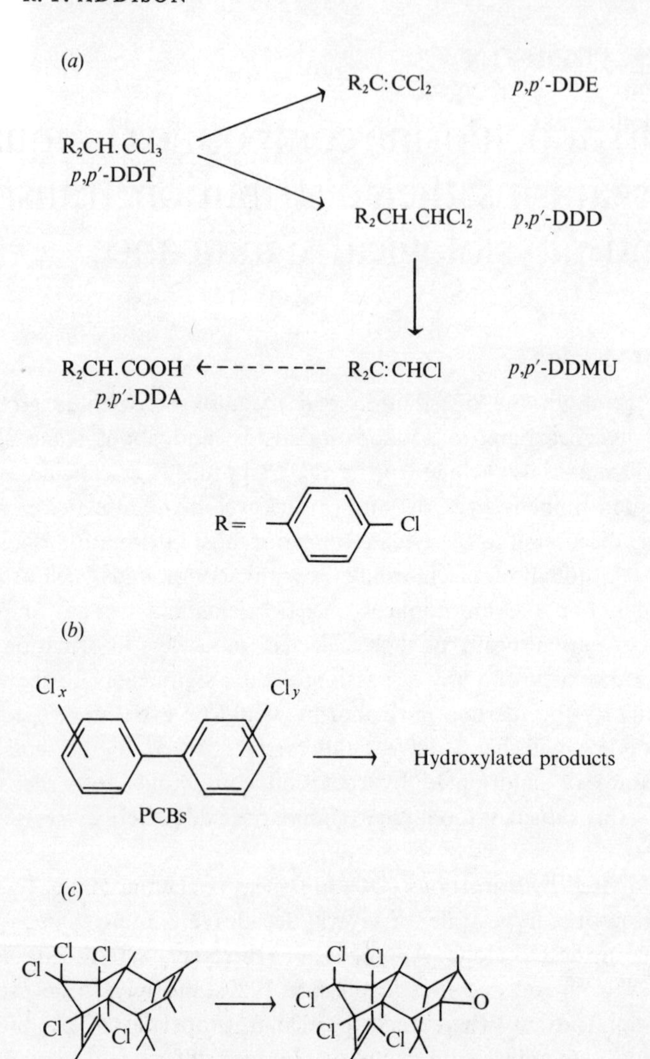

Fig. 1. Structures of some organochlorine compounds and their metabolites. (*a*) *p,p′*-DDT and some of its metabolites; (*b*) general formula for PCBs and probable metabolites; (*c*) aldrin and dieldrin.

problems in analysing the PCBs, and in interpreting their environmental significance, as they consist not of one discrete compound, but of several (some of which are still of uncertain structure) and their distribution may vary from sample to sample.

The DDT group of insecticides was introduced in the mid 1940s; *p,p′*-DDT itself is the most widely used member of the group, but several structurally related compounds, including *p,p'*-DDD, which is a metabolite of *p,p'*-DDT, also have insecticidal properties.

The cyclodiene insecticides were first extensively used in the early 1950s; although several compounds in the group are insecticidal, I shall refer to only three of them – aldrin, dieldrin and endrin. By comparison with the PCBs, both the DDT-group and the cyclodienes consist of a more limited and better-defined selection of compounds, and correspondingly more precise information is available about their behaviour in aquatic environments. Structures of representative compounds from each group are shown in Fig. 1.

All three groups are chlorinated hydrocarbons: the PCBs are derived from biphenyl, the DDT group is derived (formally) from 1,1-diphenylethane, and the cyclodienes are derived (again formally) from 1,4:5,8-dimethanonaphthalene. All groups share some common physical and chemical properties which arise partly from their hydrocarbon ancestry. The most important of these in influencing their environmental distribution are: (*a*) relative unreactivity; (*b*) appreciable volatility at normal environmental temperatures; and (*c*) insolubility in water, solubility in lipids and high lipid–water partition coefficients. Thus, since they are unreactive and hence fairly stable in various environmental reservoirs, they are persistent; they are volatile, and are transported in vapour phase, which accounts for their ubiquity, and their lipid solubility accounts for their accumulation in depot fat reserves of marine (and other) organisms.

Organochlorine residue distribution in marine organisms

The uptake and clearance of OCs by algae and aquatic invertebrates has been reviewed by Kerr & Vass (1973). They concluded that uptake of OCs by unicellular organisms could be described as adsorption onto the cell followed by absorption into the cell. (The term 'adsorption' is used frequently to describe residue accumulation by small organisms. Strictly, adsorption implies the localisation of adsorbate (in this case, OC) at the surface of the adsorbent (the cell). In practice, it is difficult to distinguish between material accumulated onto the surface of the cell and that which has penetrated within it, but there is some evidence that absorption within the cell of adsorbed material can occur (Bowes, cited by Cox, 1972). It seems reasonable to regard OC uptake by small organisms as a two-stage process, involving first adsorption to the surface, followed by absorption within the cell (Sodergren, 1968).

Kerr & Vass (1973) concluded that three processes could contribute to OC accumulation by larger and more complex invertebrates: ingestion of contaminated food, direct absorption from water through the gills and absorption through the integument of adsorbed material. The relative importance of each process would vary with environmental conditions and with the nature of the organism, but absorption through the gill and gut were probably the main routes.

Generally, OCs are eliminated by algae and invertebrates more slowly than they are accumulated. Elimination of [^{14}C]*p,p'*-DDT by *Chlorella* required about 4 days for completion (Sodergren, 1968) and small phytoplankton turned over PCBs within a few days (Ware & Addison, 1973). However, small invertebrates contaminated with [^{14}C]*p,p'*-DDT and then put in clean water did not completely eliminate the pesticide within several days (Cox, 1971). This suggests that elimination is not simply the reverse of accumulation, and in particular that it is not governed simply by the position of desorption or distribution equilibria. Either of these processes would imply that OCs accumulated by an organism should be in a dynamic equilibrium with OCs in the surrounding water. When the external OC concentration is reduced, as on substitution of clean water, there should subsequently be a continuous decrease in residue concentrations in the organisms. As noted above this does not occur, at least over short experimental periods.

Accumulation of OCs by fish can proceed by two routes: absorption through the gill (Holden, 1962; Murphy, 1971), and ingestion of contaminated food (Macek, Rodgers, Stalling & Korn, 1970; Grzenda, Paris & Taylor, 1970). Residue uptake through the gills is related to metabolic rate and body size (Murphy, 1971). In practice, food seems to be the main source of OCs to fish (Macek & Korn, 1970; Norstrom, McKinnon, deFreitas & Miller, 1975). Different OCs are absorbed through the gut with varying efficiencies: rainbow trout fed [^{14}C]*p,p'*-DDT or [^{14}C]dieldrin retained about 24 % of the former but only about 10 % of the latter (Macek *et al.*, 1970). In a separate experiment, Lieb, Bills & Sinnhuber (1974) found that rainbow trout retained about 68 % of the PCB (Aroclor 1254) fed to them. The efficiency of OC absorption through gills is also variable: bluegills and goldfish exposed briefly to ^{14}C-labelled lindane (a hexachlorocyclohexane), dieldrin or *p,p'*-DDT in water accumulated them with efficiencies increasing in that order (Gakstatter & Weiss, 1967). In general, then, the efficiency of OC accumulation by fish follows the sequence PCB > DDT > dieldrin > lindane.

OCs accumulated through the gut or gills are presumed to enter the bloodstream and several studies have been made of the clearance of OCs from blood following their intravenous injection to fish. [^{14}C]*p,p'*-DDT clearance from plasma of dogfish, skate and flounder followed a three-compartment system. The three 'compartments' were identified as: first, clearance from plasma; second, redistribution from tissues of high blood flow in which DDT was stored temporarily after plasma clearance; and third, re-establishment of an equilibrium between plasma and tissue DDT levels (Dvorchik &

Maren, 1972; Darrow & Addison, 1973; Pritchard, Guarino & Kinter, 1973). Half-lives of [^{14}C]*p,p*′-DDT in the three 'compartments' ranged from a few minutes in the first to several hours or more in the last. Even with the same species OCs were handled in different ways: clearance of the insecticide Mirex from flounder plasma differed from that of *p,p*′-DDT (Pritchard *et al.*, 1973) and simultaneous injection of 2-, 3- and 4-chlorobiphenyl to skate resulted in a faster plasma clearance of the 3-isomer than of the other two (Zinck & Addison, 1974).

Elimination of OCs by fish proceeds more slowly than does their accumulation and there are some consistent differences between compounds. [^{14}C]dieldrin accumulated over a prolonged feeding period was eliminated more rapidly from rainbow trout and goldfish than was [^{14}C]*p,p*′-DDT acquired similarly (Macek *et al.*, 1970; Grzenda *et al.*, 1970). It is difficult to compare directly elimination rates of OCs accumulated from food with those of OCs accumulated from water (Macek *et al.*, 1970) since the latter experiments have usually involved shorter periods of OC exposure, and the OCs may be distributed differently from those in fed fish. Nevertheless, ^{14}C-labelled lindane, dieldrin and *p,p*′-DDT accumulated from water by goldfish and bluegills were eliminated at rates decreasing in that order (Gakstatter & Weiss, 1967). PCBs (Clophen A-50) accumulated from water by goldfish were cleared with a half-life of about 15–20 days (Hattula & Karlog, 1973) and paralleled a reduction in whole-body fat levels. (There are difficulties in comparing these data with others since they were obtained from smaller fish maintained at a higher temperature than the goldfish used by Gakstatter & Weiss (1967). These differences suggest a higher overall metabolic rate for Hattula & Karlog's sample and since Gakstatter & Weiss reported a half-life of over 32 days for *p,p*′-DDT it seems that the rate of elimination of PCBs is of the same order as that of *p,p*′-DDT.) To generalise from these data, the tendency to eliminate OCs follows the sequence lindane $>$ dieldrin $>$ DDT $\geqslant$ PCB.

There is no evidence to suggest that metabolism of OCs is a necessary prelude to their elimination. Grzenda *et al.* (1970), found that appreciable amounts of the [^{14}C]*p,p*′-DDT which they fed to their goldfish were metabolised to *p,p*′-DDD and *p,p*′-DDE. Faeces contained both *p,p*′-DDT and *p,p*′-DDE but (rather surprisingly) no *p,p*′-DDD which would be expected to have been formed by intestinal micro-organisms (discussed further below).

We know relatively little about the accumulation of OCs by marine mammals. Frank, Ronald & Braun (1973) showed that captive seals accumulated residues of DDT, PCBs and cyclodienes from their diet, but they did not present sufficient data to allow any estimate of assimilation efficiency to be made. Indirect evidence (discussed in more detail below) suggests that clearance of OCs from mammals is relatively low.

Since OC clearance rates are generally slower than uptake rates, the residue burden of an organism increases with time (assuming a constant rate of exposure). This is reflected in an increase in OC residue concentrations with age, and such relationships have been demonstrated for the DDT-group and PCBs in freshwater fish (Youngs, Gutenmann & Lisk, 1972; Bache, Serum, Youngs & Lisk, 1972), marine fish (Jensen, Johnels, Olsson & Otterlind, 1972) and seals (Addison, Kerr, Dale & Sergeant, 1973; Addison & Smith, 1974); and for dieldrin in sea birds (Robinson *et al.*, 1967). The form of the relationship may vary: Youngs *et al.* (1972) and Bache *et al.* (1972) found that OC residue concentrations were exponentially related to age in lake trout, whereas in the seal samples we have examined linear relationships fitted the data best. Such differences may arise from several sources; one of them may be the assumption that exposure to OCs over prolonged periods is constant, which (in some cases) it is not (Kerswill, 1967).

OC residues are stored mainly in an organisms' depot fat and their distribution may depend to some extent on how the organism uses that fat. Two extreme cases can be envisaged: in one, residues stored in depot fat are handled completely independently of the fat; in the other case, mobilisation of the fat is unselective in the sense that any residues deposited in fat are mobilised along with it. An example of the latter case is the mobilisation of blubber lipid by the female seal; we have found that OC concentrations in female blubber do not increase with the age of the animal (in contrast to those in male seals) which suggests that the female mobilises and secretes residues along with fat during lactation (Addison & Smith, 1974). Analyses of seal milk support this view (Anas & Wilson, 1970; R. F. Addison & P. F. Brodie, unpublished data). The other extreme is illustrated by analyses of OCs in Atlantic herring, where whole-body OC concentrations remain approximately constant despite wide seasonal fluctuations in the fat content of the fish (Addison, Zinck & Ackman, 1972). This is not necessarily a general feature of fat mobilisation in fish, however: Earnest & Benville (1971) have found both positive and negative correlations between OC concentration and fat content (both expressed on a whole-body basis) from several species of fish.

To summarise the results describing OC uptake, distribution and elimination: OCs can be accumulated through adsorption to external surfaces, absorption from water through gills or other absorptive tissue, and by ingestion. The importance of each route varies with the organism, adsorption predominating in small invertebrates and ingestion in large vertebrates. The extent to which OCs are retained by organisms depends on the structure of the OC (probably increasing in the sequence lindane $<$ dieldrin $<$ DDT $\leqslant$ PCB) and on biological factors, including the processes of fat mobilisation in the organism.

Metabolism of organochlorines by aquatic organisms

The foregoing discussion has treated the OCs as if they were stored and excreted from organisms without undergoing any metabolism. However, many aquatic organisms metabolise OCs, though apparently more slowly than do terrestrial organisms. Before examining the capacity of marine organisms to metabolise OCs, I shall describe briefly the main pathways of OC degradation established by work on other animals. Several comprehensive reviews are available elsewhere (O'Brien, 1967; Menzie, 1969; Fukuto & Sims, 1971).

Degradation of *p,p'*-DDT can follow several routes. One leads to *p,p'*-DDE, which is probably a metabolic dead-end; another goes to a hydroxylated derivative, dicofol (at least in some insects). The main excretory route in mammals probably passes through *p,p'*-DDD via dehydrochlorinations, reductions and oxidations to *p,p'*-DDA and similar polar metabolites. Among the cyclodienes, aldrin is generally epoxidised to dieldrin which, in turn, may be further metabolised to hydroxylated derivatives, at least by mammals. The metabolism of PCBs varies with the structure of the compound but generally leads to the production of mono- or di-hydroxy chlorobiphenyls (Hutzinger *et al.*, 1974). Some of these pathways are summarised in Fig. 1.

Aquatic micro-organisms are able to carry out some of the conversions listed above. Patil, Matsumura & Boush (1972) concluded that thirty-five out of 100 microbial isolates from a miscellany of sources, including open ocean water, surface films and sediments, could degrade [^{14}C]*p,p'*-DDT, the main product being *p,p'*-DDD. (This is the usual metabolic product of *p,p'*-DDT degradation by micro-organisms.) Examination of their data (Patil *et al.*, 1972, Tables 2 and 3) shows that large amounts of labelled material apparently remained in the aqueous phases after solvent extraction, which suggests that water-soluble metabolites were also formed. Several of their algal and plankton samples could also convert [^{14}C]*p,p'*-DDT to *p,p'*-DDD and [^{14}C]aldrin to dieldrin, or [^{14}C]dieldrin to various metabolites. In other studies, phytoplankton organisms converted [^{14}C]*p,p'*-DDT to *p,p'*-DDE; rates of production ranged from negligibly slow to about 7.5 % conversion over a 9-day experimental period (Kiel & Priester, 1969; Bowes, 1972; Rice & Sikka, 1973).

Some micro-organisms can metabolise 4-chlorobiphenyl by hydroxylation, or by cleavage of the unsubstituted ring to form 4-chlorobenzoic acid (Ahmed & Focht, 1973; Hutzinger *et al.*, 1974). There is as yet no information about the capacity of specifically marine micro-organisms to perform such degradations.

Among other invertebrates, planarian worms converted *p,p'*-DDT mainly to *p,p'*-DDD and some *p,p'*-DDE (Phillips, Wells & Chandler, 1974). Since these metabolites were detected after *p,p'*-DDT was fed to the worms it is possible that they were produced by the animals' intestinal flora. Somewhat unexpectedly, copepods (*Calanus* spp.) could not metabolise [^{14}C]*p,p'*-DDT accumulated from water even after 8 weeks' exposure. This may have been due to its being localised at the external surfaces of the animals, where it may have been inaccessible to the appropriate enzymes (Darrow & Harding, 1975). Lobsters metabolised an intravascular dose of [^{14}C]*p,p'*-DDT to *p,p'*-DDD and *p,p'*-DDE (in approximately equal proportions) and (in small amounts) to *p,p'*-DDA (Guarino, Pritchard, Anderson & Rall, 1974); they also epoxidised aldrin to dieldrin (Carlson, 1974).

Considerable information exists about the ability of fish to metabolise OC compounds. Data from analyses of 'wild' samples suggest that they can convert *p,p'*-DDT to *p,p'*-DDE and probably to *p,p'*-DDD also, and that they can epoxidise aldrin to dieldrin. Since these results have been reviewed recently by Johnson (1973) the present discussion is restricted to experimental studies of OC metabolism by fish.

Several studies in which *p,p'*-DDT was fed, or exposed in the water, to fish have shown that it is converted to *p,p'*-DDE and to *p,p'*-DDD (e.g. Greer & Paim, 1968; Young, St John & Lisk, 1971). In feeding experiments it is difficult to eliminate the possibility that some of the conversions are carried out by intestinal micro-organisms, which are particularly active in producing *p,p'*-DDD (Wedemeyer, 1968; Cherrington, Paim & Page, 1969). Following injection of [^{14}C]*p,p'*-DDT to winter flounder, label was detected in *p,p'*-DDD, *p,p'*-DDE, *p,p'*-DDMU and *p,p'*-DDA (Pritchard *et al.*, 1973). We have examined the possibility that metabolism of *p,p'*-DDT by fish beyond the *p,p'*-DDD stage is similar to that in mammals, by injecting trout with ^{14}C-labelled *p,p'*-DDT, *p,p'*-DDE, *p,p'*-DDD or *p,p'*-DDMU and analysing the metabolites at intervals up to 5 weeks after injection. *p,p'*-DDT was converted to *p,p'*-DDE (about 10 % in 5 weeks) and *p,p'*-DDD (1 %); *p,p'*-DDD was converted to *p,p'*-DDMU (less than 1 %); *p,p'*-DDE and *p,p'*-DDMU did not undergo any change (Addison & Zinck, 1974). The stability of *p,p'*-DDMU was somewhat unexpected, since by analogy with the mammalian route of *p,p'*-DDT metabolism, we anticipated reduction to *p,p'*-DDMS. Taken together, all these results suggest that *p,p'*-DDT metabolism in fish may follow the same route as that in mammals, at least in its initial stages, but that the rates of metabolism are appreciably slower. This may be partly due to the lower environmental temperatures of most fish studied; we found, for example, that the rate of *p,p'*-DDT to *p,p'*-DDE conversion more than doubled over the interval 2–18 °C (Zinck & Addison, 1975). The very slow production of *p,p'*-DDD relative to that of *p,p'*-DDE is consistent with the distribution of DDT metabolites observed in 'wild' samples, where *p,p'*-DDD is generally a minor component.

Experimental studies have also shown that fish can epoxidise aldrin to dieldrin (Ludke, Gibson & Lusk, 1972).

The metabolism of PCBs by fish has not yet been studied in any detail, probably because of the shortage of individual PCB components until fairly recently. When fish accumulate PCBs from water or food, there are generally some differences between peak distribution in the standard and that in the accumulated material (e.g. Hansen *et al.*, 1971; Hattula & Karlog, 1973). However, these differences are usually detected after brief exposure to PCBs and are attributable to selectivity in uptake or elimination of certain components of the mixture, rather than to metabolic degradative processes.

The ability to fish to metabolise biphenyl has been studied. This is a convenient model compound for exploring possible routes of PCB metabolism and is also a major component of the simpler PCB mixtures (Willis & Addison, 1972). Biphenyl can be hydroxylated in the 4-position by trout liver preparations *in vitro* (Creaven, Parke & Williams, 1965) and in the 2- and 4-positions by skate liver preparations (Willis & Addison, 1974). Similar hydroxylations proceed slowly *in vivo* in the skate; less than 1 % of an intravenously administered dose of biphenyl was hydroxylated within 5 days (D. E. Willis & R. F. Addison, unpublished data). Most of the metabolites were in the bile.

The metabolism of individual chlorobiphenyls by fish was investigated by Hutzinger *et al.* (1972). They were unable to detect any hydroxylated products by analysing the water in which the fish were kept, but in the light of our experience of the slow hydroxylation of biphenyl, these results are not surprising.

The metabolism of OCs by marine mammals has not yet been studied experimentally in any detail. In an analysis of faeces collected from a captive seal in a zoo, Jansson *et al.* (1975) found two phenolic *p,p'*-DDE derivatives. These were present in the bile of other samples suggesting that they were not produced by the animal's intestinal flora. Similar phenolic metabolites of PCBs were detected. The distribution of DDT-group compounds in seal blubber changes with age, the proportion existing as *p,p'*-DDE increasing. This result is consistent with the animal's expected ability to convert *p,p'*-DDT to *p,p'*-DDE (Addison *et al.*, 1973).

Discussion

Two views have been advanced to explain the distributions of OC compounds observed in 'environmental' samples. One of these (Woodwell, Wurster & Isaacson, 1967) was proposed to explain residue distributions observed in organisms living in a salt-marsh; it holds that an organism's residue burden depends on the amount of residue available from components in its diet. On this scheme, an organism's residue burden should depend on its position in the food web and, in that sense, be governed by biological factors. An alternative view (due to Hamelink, Waybrant & Ball, 1971) was put forward to explain DDT distributions in organisms in ponds after experimental addition of DDT; it proposes that OC levels in an organism depend mainly on the physical properties of the OC in question, and particularly on those properties which control its tendency to distribute itself between lipoid and aqueous phases. However, the evidence from experimental and field studies, summarised above, suggests that in many cases it is unreasonable to try to explain OC distributions on the basis of either one or other of these approaches; instead, each view can contribute something towards explaining the residue distribution patterns observed.

In larger invertebrates, and in vertebrates, the experimental evidence suggests that food is the major OC residue source under normal circumstances. Since clearance of residues in these (and other) organisms is much slower than residue uptake, residue burdens increase with time and there should be a selective retention of OCs at increasing trophic levels. One result of this is that the OC concentrations (on a whole-body basis) should increase with trophic level. This is more or less what Woodwell *et al.* (1967) observed; in field studies where no increase in residue concentration with trophic level was observed (e.g. Peterle, 1966) most of the organisms studied came from lower trophic levels – where residues may be accumulated from water rather than food.

The 'partition' theory succeeds in explaining experimental observations of residue uptake directly from water, and especially those in which structural differences in residue are reflected in differences in their retention or elimination (e.g. Gakstatter & Weiss, 1967). Those results which show that residue uptake is dose-dependent are also consistent with the 'partition' hypothesis (e.g. Macek *et al.*, 1970). However, most of the data on which the proposal of Hamelink *et al.* (1971) was based came from relatively short-term studies involving OC accumulation from water and, although it explained their (and others') observations, it may not account for residue concentrations in predator organisms where uptake arises mainly from food. But where a simple partition theory fails most conspicuously is in explaining residue clearance from organisms. A lipid–water partition hypothesis requires that, given clean water, residues will eventually disappear completely from an organism. This is not borne out by most experimental evidence, at least for higher organisms (e.g. Macek *et al.*, 1970), although some data (Grzenda, Taylor & Paris, 1972) do support that hypothesis. The observations that OC concentrations increase with age in various animals also refutes this concept, though not entirely convincingly in the absence of OC analysis of the water from which the samples were taken. However, most data suggest that clearance of residue from organisms depends not so much on the position of a lipid–water distribution equilibrium, but on the time taken to attain it.

To summarise, in situations where an organism accumulates the bulk of its OC burden by direct contact with water, the proposal that the amount taken up is controlled by some factor closely related to the lipid–water partition coefficient of the OC provides a semi-quantitative description of the net uptake processes. Where an organism accumulates most of its residue burden from food, whole-body residue concentrations should depend to some extent on the trophic level it occupies, but even here the efficiency of retention of OCs is apparently governed by some factors similar to those governing their lipid–water distribution.

The significance of organochlorines to marine organisms

OCs have been shown to produce several effects on aquatic organisms: aqueous suspensions of OC insecticides are toxic to aquatic invertebrates and fish (Kerr & Vass, 1973; Holden, 1973) though acute toxicity generally requires concentrations at least an order of magnitude higher than 'natural' concentrations (e.g. Edwards, 1970). Sublethal aqueous concentrations produce changes in temperature selection by fish (Gardner, 1973), learning by fish (Hatfield & Johansen, 1972) and in predation by one fish on another (Hatfield & Anderson, 1972). OCs also cause specific biochemical changes; the best-documented of these is interference with ATPases, which may occur at tissue OC concentrations not far above those routinely encountered (Leadem, Campbell & Johnson, 1974). In general, it is difficult to assess the practical significance of many such experimental observations: the crucial question is whether 'natural' tissue or water OC concentrations produce any deleterious effects.

One effect of OCs which seems particularly important is their capacity to induce the enzymes required for their degradation. OCs (and several other foreign compounds) are metabolised by a rather non-specific group of microsomal enzymes, the mixed function oxidases (MFOs). On exposure of the organism to OCs (among several types of inducers) MFO activity increases and the organism's capacity to degrade the inducer (and other foreign compounds) is enhanced. MFO induction by OCs has been demonstrated in terrestrial invertebrates and vertebrates (e.g. Chhabra & Fouts, 1974), where it is considered to be a protective mechanism, by accelerating the conversion of a toxic OC to a non-toxic derivative, or the conversion of a relatively non-polar OC to a relatively polar (and hence water-soluble) excretory product.

There is as yet no direct evidence that MFO systems in aquatic organisms are inducible by OCs. MFOs exist in such organisms (Ludke *et al.*, 1972; Stanton & Khan, 1973; Carlson, 1974) and some indirect evidence suggests that the system may be inducible. Thus Grzenda *et al.* (1970) found that *p,p'*-DDT metabolism was accelerated on prolonged exposure to *p,p'*-DDT, and Mayer, Street & Neuhold (1970) have described various interactions of OCs in fish similar to those found in induced MFO systems in other organisms.

Even if aquatic organisms do possess an MFO system, inducible by exposure to OCs in the same way as that which occurs in terrestrial mammals, we still have no idea as to what its significance to the organism should be. Is degradation of OCs to relatively polar products necessary for their excretion, and will prolonged exposure to OCs (or to other inducers) stimulate elimination by accelerating the rate at which such conversions occur? In this context, it is interesting to consider the phenomenon of insecticide-resistant fish: prolonged 'pressure' from cyclodiene insecticides (mainly endrin and dieldrin), mainly in the southern states of the US, has resulted in the selection of an insecticide-resistant strain (e.g. Dzuik & Plapp, 1973). These fish have high MFO activity (Wells, Ludke & Yarborough, 1973), but whether this is the characteristic which confers resistance or whether it is a result of the high tissue OC concentrations which they tolerate is not yet clear.

Closing remarks. In spite of its all-encompassing title, this review has been unbalanced. I have emphasised some aspects of the subject which reflect my own interests, but some of the imbalance reflects the state of our knowledge of the interactions between OCs and aquatic organisms. Intensive environmental monitoring operations have been carried out for more than a decade, and from them (and from the experimental work which has complemented them) we can infer a good deal about the distribution of OCs in organisms and about the environmental and biological factors which affect their distribution. Generally, this information describes the relationships between OCs within an organism to those in its external environment. We are, however, less able to assess the significance of OC residue levels within organisms: what does a concentration of x ppm *p,p'*-DDT in depot fat of a fish mean, in terms of physiological or biochemical changes? Some situations exist, of course, where a given residue concentration is associated with a clear response, such as death (e.g. Dacre & Scott, 1971) but these examples are infrequent: many organisms carry an (apparently) sublethal burden of OCs sequestered in depot fat, whose potentially deleterious effects we cannot yet assess. In future we should devote more effort to examining the distribution of OCs in the organism's internal environment; a crucial problem in this area is to establish the relationship of OCs sequestered in depot fat to their effects on 'target' systems. With this information, any future review of the subject could deal more with the effects of OCs on aquatic organisms, than with the effect of organisms on OCs.

References

Addison, R. F., Kerr, S. R., Dale, J. & Sergeant, D. E. (1973). Variation in organochlorine residue levels with age in Gulf of St Lawrence harp seals (*Pagophilus groenlandicus*). *Journal of the Fisheries Research Board of Canada*, **30**, 595–600.

Addison, R. F. & Smith, T. G. (1974). Organochlorine residue levels in Arctic ringed seals: variation with age and sex. *Oikos*, **25**, 335–7.

Addison, R. F. & Zinck, M. E. (1974). The metabolism of some DDT-type compounds by Brook Trout (*Salvelinus fontinalis*). *Environmental Quality and Safety*, in press.

Addison, R. F., Zinck, M. E. & Ackman, R. G. (1972). Residues of organochlorine pesticides and polychlorinated biphenyls in some commercially produced Canadian marine oils. *Journal of the Fisheries Research Board of Canada*, **29**, 349–55.

Ahmed, M. & Focht, D. D. (1973). Degradation of polychlorinated biphenyls by two species of *Achromobacter*. *Canadian Journal of Microbiology*, **19**, 47–52.

Anas, R. E. E. & Wilson, A. J. (1970). Organochlorine pesticides in nursing fur seal pups. *Pesticides Monitoring Journal*, **4**, 114–16.

Anon. (1969). *Cleaning our environment: the chemical basis for action*. American Chemical Society, Washington.

Bache, C. A., Serum, J. W., Youngs, W. D. & Lisk, D. J. (1972). Polychlorinated biphenyl residues: accumulation in Cayuga Lake trout with age. *Science*, **177**, 1191–2.

Bowes, G. W. (1972). Uptake and metabolism of 2,2-bis-(*p*-chlorophenyl)-1,1,1-trichloroethane (DDT) by marine phytoplankton and its effect on growth and chloroplast electron transport. *Plant Physiology*, **49**, 172–6.

Carlson, G. P. (1974). Epoxidation of aldrin to dieldrin by lobsters. *Bulletin of Environmental Contamination and Toxicology*, **11**, 577–82.

Cherrington, A. D., Paim, U. & Page, O. T. (1969). *In vitro* degradation of DDT by intestinal contents of Atlantic salmon (*Salmo salar*). *Journal of the Fisheries Research Board of Canada*, **26**, 47–54.

Chhabra, R. S. & Fouts, J. R. (1974). Stimulation of hepatic drug-metabolising enzymes by DDT polycyclic hydrocarbons or phenobarbitol in adrenalectomised or castrated mice. *Toxicology and Applied Pharmacology*, **28**, 465–76.

Cox, J. L. (1971). Uptake, assimilation and loss of DDT residues by *Euphausia pacifica*, a euphausiid shrimp. *Fishery Bulletin*, **69**, 627–33.

Cox, J. L. (1972). DDT residues in marine phytoplankton. *Residue Reviews*, **44**, 23–38.

Creaven, P. J., Parke, D. V. & Williams, R. T. (1965). A fluorometric study of the hydroxylation of biphenyl *in vitro* by liver preparations of various species. *Biochemical Journal*, **96**, 879–85.

Dacre, J. C. & Scott, D. (1971). Possible DDT mortality in young rainbow trout. *New Zealand Journal of Marine and Freshwater Research*, **5**, 58–65.

Darrow, D. C. & Addison, R. F. (1973). The metabolic clearance of ^{14}C-*p,p'*-DDT from plasma, and its distribution in the thorny skate, *Raja radiata*. *Environmental Physiology and Biochemistry*, **3**, 196–203.

Darrow, D. C. & Harding, G. C. (1975). Accumulation and apparent absence of metabolism of DDT by marine copepods, *Calanus* spp. *Journal of the Fisheries Research Board of Canada*, **32**, 1845–9.

Dvorchik, B. H. & Maren, T. H. (1972). The fate of *p,p'*-DDT (2,2-bis-(*p*-chlorophenyl)-1,1,1-trichloroethane) in the dogfish, *Squalus acanthias*. *Comparative Biochemistry and Physiology*, **42A**, 205–11.

Dzuik, L. J. & Plapp, F. W. (1973). Insecticide resistance in mosquito fish from Texas. *Bulletin of Environmental Contamination and Toxicology*, **9**, 15–19.

Earnest, R. D. & Benville, P. E. (1971). Correlation of DDT and lipid levels for certain San Francisco Bay Fish. *Pesticides Monitoring Journal*, **5**, 235–41.

Edwards, C. A. (1970). Persistent pesticides in the environment. *Critical Reviews in Environmental Control*, **1**, 7–67.

Frank, R., Ronald, K. & Braun, H. E. (1973). Organochlorine residues in harp seals (*Pagophilus groenlandicus*) caught in Eastern Canadian waters. *Journal of the Fisheries Research Board of Canada*, **30**, 1053–63.

Fukuto, T. R. & Sims, J. J. (1971). *Metabolism of insecticides and fungicides*. In *Pesticides in the environment*, vol. 1, part 1, (ed. R. White-Stevens), pp. 145–236. Marcel Dekker, New York.

Gakstatter, J. H. & Weiss, C. M. (1967). The elimination of DDT-^{14}C, dicldrin-^{14}C and lindane-^{14}C from fish following a single sub-lethal dose in aquaria. *Transactions of the American Fisheries Society*, **96**, 301–7.

Gardner, D. R. (1973). The effect of some DDT and methoxychlor analogs on temperature selection and lethality in brook trout fingerlings. *Pesticide Biochemistry and Physiology*, **2**, 437–46.

Greer, G. L. & Paim, U. (1968). Degradation of DDT in Atlantic salmon (*Salmo salar*). *Journal of the Fisheries Research Board of Canada*, **25**, 2321–6.

Grzenda, A. R., Paris, D. F. & Taylor, W. J. (1970). The uptake, metabolism and elimination of chlorinated residues by goldfish (*Carassius auratus*) fed a ^{14}C-DDT contaminated diet. *Transactions of the American Fisheries Society*, **99**, 385–96.

Grzenda, A. R., Taylor, W. J. & Paris, D. F. (1972). The elimination and turnover of ^{14}C-dieldrin by different goldfish tissue. *Transactions of the American Fisheries Society*, **101**, 686–90.

Guarino, A. M., Pritchard, J. B., Anderson, J. B. & Rall, D. B. (1974). Tissue distribution of (^{14}C) DDT in the lobster after administration by intravascular or oral routes, or after exposure from ambient seawater. *Toxicology and Applied Pharmacology*, **29**, 277–88.

Hamelink, J. L., Waybrant, R. C. & Ball, R. C. (1971). A proposal: exchange equilibria control the degree chlorinated hydrocarbons are biologically magnified in lentic environments. *Transactions of the American Fisheries Society*, **100**, 207–14.

Hansen, D. J., Parrish, P. R., Lowe, J. I., Wilson, A. J. & Wilson, P. D. (1971). Chronic toxicity, uptake and retention of Aroclor[R] 1254 in two estuarine fishes. *Bulletin of Environmental Contamination and Toxicology*, **6**, 113–19.

Hatfield, C. T. & Anderson, J. M. (1972). Effects of two insecticides on the vulnerability of Atlantic salmon (*Salmo salar*) parr to brook trout (*Salvelinus fontinalis*) predation. *Journal of the Fisheries Research Board of Canada*, **29**, 27–9.

Hatfield, C. T. & Johansen, P. H. (1972). Effects of four insecticides on the ability of Atlantic salmon parr (*Salmo salar*) to learn and retain a simple conditioned response. *Journal of the Fisheries Research Board of Canada*, **29**, 315–21.

Hattula, M. L. & Karlog, O. (1973). Absorption and elimination of polychlorinated biphenyls (PCB) in goldfish. *Acta Pharmalogica et Toxicologica*, **32**, 237–45.

Holden, A. V. (1962). A study of the absorption of ^{14}C-labelled DDT from water by fish. *Annals of Applied Biology*, **50**, 467–77.

Holden, A. V. (1973). Effects of pesticides on fish. In *Environmental pollution by pesticides*, (ed. C. A. Edwards), pp. 213–53. Plenum Press, London.

Hutzinger, O., Nash, D. M., Safe, S. H., deFreitas, A. S. W., Norstrom, R. J., Wildish, D. J. & Zitko, V. (1972). Polychlorinated biphenyls: metabolic behaviour of pure isomers in pigeons, rats and brook trout. *Science*, **178**, 312–16.

Hutzinger, O., Safe, S. H. & Zitko, V. (1974). *The chemistry of PCB's*. Chemical Rubber Co. Press, Cleveland, Ohio.

Jansson, B., Jensen, S., Olsson, M., Renberg, L., Sundstrom, G. & Vaz, R. (1975). Identification by GC-MS of phenolic metabolites of PCB and *p,p'*-DDE isolated from Baltic guillemot and seal. *Ambio*, in press.

Jensen, S., Johnels, A. G., Olsson, M. & Otterlind, G. (1972). DDT and PCB in herring and cod from the Baltic, the Kattegat and the Skagerrak. *Ambio Special Report*, **1**, 71–85.

Johnson, D. W. (1973). Pesticide residues in fish. In *Environmental pollution by pesticides*, (ed. C. A. Edwards), pp. 181–212. Plenum Press, London.

Kiel, J. E. & Priester, L. A. (1969). DDT uptake and metabolism by a marine diatom. *Bulletin of Environmental Contamination and Toxicology*, **4**, 169–73.

Kerr, S. R. & Vass, W. P. (1973). Pesticide residues in aquatic invertebrates. In *Environmental pollution by pesticides*, (ed. C. A. Edwards), pp. 134–80 Plenum Press, London.

Kerswill, C. J. (1967). Studies on effects of forest spraying with insecticides, 1952–1963, on fish and aquatic invertebrates in New Brunswick streams: introduction and summary. *Journal of the Fisheries Research Board of Canada*, **24**, 701–8.

Leadem, T. P., Campbell, R. D. & Johnson, D. W. (1974). Osmoregulatory responses to DDT and varying salinities in *Salmo gairdneri*. I. Gill Na-K-ATPase. *Comparative Biochemistry and Physiology*, **49A**, 197–205.

Lieb, A. F., Bills, D. D. & Sinnhuber, R. O. (1974). Accumulation of dietary polychlorinated biphenyls (Aroclor 1254) by rainbow trout (*Salmo gairdneri*). *Journal of Agricultural and Food Chemistry*, **22**, 638–42.

Ludke, J. L., Gibson, J. R. & Lusk, C. I. (1972). Mixed function oxidase activity in freshwater fishes: aldrin epoxidation and parathion activation. *Toxicology and Applied Pharmacology*, **21**, 89–97.

Macek, K. J. & Korn, S. (1970). Significance of the food chain in DDT accumulation by fish. *Journal of the Fisheries Research Board of Canada*, **27**, 1496–8.

Macek, K. J., Rodgers, C. R., Stalling, D. L. & Korn, S. (1970). The uptake, distribution and elimination of dietary ^{14}C-DDT and ^{14}C-dieldrin in rainbow trout. *Transactions of the American Fisheries Society*, **99**, 689–95.

Mayer, F. L., Street, J. C. & Neuhold, J. M. (1970). Organochlorine insecticide interactions affecting residue storage in rainbow trout. *Bulletin of Environmental Contamination and Toxicology*, **5**, 300–10.

Menzie, C. M. (1969). *Metabolism of pesticides. Special scientific report: wildlife no. 127*. US Department of the Interior, Washington, DC.

Murphy, P. G. (1971). The effect of size on the uptake of DDT from water by fish. *Bulletin of Environmental Contamination and Toxicology*, **6**, 20–3.

Norstrom, R. J., McKinnon, A. E., deFreitas, A. S. W. & Miller, D. R. (1975). Pathway definition of pesticide and mercury uptake by fish. *Environmental Quality and Safety*, in press.

O'Brien, R. D. (1967). *Insecticides: action and metabolism*. Academic Press, New York & London.

Patil, K. C., Matsumura, F. & Boush, G. M. (1972). Metabolic transformation of DDT, dieldrin, aldrin and endrin by marine micro-organisms. *Environmental Science and Technology*, **6**, 629–32.

Peterle, T. J. (1966). The use of isotopes to study pesticide translocation in natural environments. *Journal of Applied Ecology*, **3**, Supplement, 181–90.

Phillips, J., Wells, M. & Chandler, C. (1974). Metabolism of DDT by the freshwater planarian, *Phagocata velata*. *Bulletin of Environmental Contamination and Toxicology*, **12**, 355–7.

Pritchard, J. B., Guarino, A. M. & Kinter, W. B. (1973). Distribution, metabolism and excretion of DDT and mirex by a marine teleost, the winter flounder. *Environmental Health Perspectives*, **4**, 45–54.

Rice, C. P. & Sikka, H. (1973). Uptake and metabolism of DDT by six species of marine algae. *Journal of Agricultural and Food Chemistry*, **21**, 148–52.

Robinson, J., Richardson, A., Crabtree, A. N., Coulson, J. C. & Potts, G. R. (1967). Organochlorine residues in marine organisms. *Nature, London*, **214**, 1307–11.

Sodergren, A. (1968). Uptake and accumulation of ^{14}C-DDT by *Chlorella* sp. (Chlorophyceae). *Oikos*, **19**, 126–38.

Stanton, R. H. & Khan, M. A. Q. (1973). Mixed function oxidase activity towards cyclodiene insecticides in bass and bluegill sunfish. *Pesticide Biochemistry and Physiology*, **3**, 351–7.

Ware, D. M. & Addison, R. F. (1973). PCB residues in plankton from the Gulf of St Lawrence. *Nature, London*, **216**, 519–21.

Wedemeyer, G. (1968). Role of intestinal microflora in the degradation of DDT by rainbow trout (*Salmo gairdneri*). *Life Sciences*, **7**, 219–23.

Wells, M. R., Ludke, J. L. & Yarborough, J. D. (1973). Expoxidation and fate of (^{14}C)aldrin in insecticide-resistant and susceptible populations of mosquito fish (*Gambusia affinis*). *Journal of Agricultural and Food Chemistry*, **21**, 428–9.

Willis, D. E. & Addison, R. F. (1972). Identification and estimation of the major components of a commercial polychlorinated biphenyl mixture, Aroclor 1221. *Journal of the Fisheries Research Board of Canada*, **29**, 592–5.

Willis, D. E. & Addison, R. F. (1974). Hydroxylation of biphenyl *in vitro* by tissue preparations of some marine organisms. *Comparative and General Pharmacology*, **5**, 77–81.

Woodwell, G. M., Wurster, C. F. & Isaacson, P. D. (1967). DDT residues in an East Coast Estuary: a case of biological concentration by a persistent pesticide. *Science*, **156**, 821–3.

Young, R. G., St John, L. & Lisk, D. J. (1971). Degradation of DDT by goldfish. *Bulletin of Environmental Contamination and Toxicology*, **6**, 357–9.

Youngs, W. D., Gutenmann, W. H. & Lisk, D. J. (1972). Residues of DDT in lake trout as a function of age. *Environmental Science and Technology*, **6**, 451–2.

Zinck, M. E. & Addison, R. F. (1974). The fate of 2-, 3-, and 4-chlorobiphenyl following intravenous administration to the thorny skate (*Raja radiata*) and the winter skate (*Raja ocellata*). *Archives of Environmental Contamination and Toxicology*, **2**, 52–62.

Zinck, M. E. & Addison, R. F. (1975). The effect of temperature on the rate of conversion of *p,p'*-DDT to *p,p'*-DDE in brook trout, *Salvelinus fontinalis*. *Canadian Journal of Biochemistry*, **53**, 636–9.

M.R.REEVE, G.D.GRICE, V.R.GIBSON,
M.A.WALTER, K.DARCY & T.IKEDA

A controlled environmental pollution experiment (CEPEX) and its usefulness in the study of larger marine zooplankton under toxic stress

Introduction

The complex interactions between communities and their physical–chemical environment are important factors in maintaining the stability of marine ecosystems. Considerable variation occurs naturally in population size and composition, which is usually unpredictable and the causes for it not generally understood. It would require many years of environmental monitoring to isolate pollution-related fluctuations in natural populations from those caused by other factors, except in catastrophic situations. In order to understand and, more importantly, to predict the effect of pollutants on marine ecosystems, it is necessary to enclose or capture the ecosystem, perturbate it with known amounts of a pollutant, and compare it with an unpolluted ecosystem by making measurements on those processes which are most likely to influence population stability. A proposal was put to the National Science Foundation that a series of such experiments be undertaken as part of the contribution to the International Decade of Ocean Exploration. The proposal was accepted and the multidisciplinary programme CEPEX (Controlled Ecosystem Pollution Experiment), involving scientists from the United States, Canada and Britain, was initiated under the overall guidance of Dr David W. Menzel of Skidaway Institute of Oceanography, Georgia. CEPEX is based on the concept of a series of plastic, transparent flexible containers into each of which can be dosed selected pollutants at levels no more than an order of magnitude above present maximum environmental levels. The aim is to examine population changes in the individual controlled experimental ecosystems (CEEs) of trophic levels ranging from micro-organisms to small fish as a basis for modelling and extrapolating to the marine ecosystem as a whole.

The CEEs should also provide a means of identifying short-term sensitive indicators of long-term sublethal pollution by correlating subtle effects on population structure and productivity with metabolic, behavioural or other quickly measurable parameters of stress. Such indicators could then be developed into future, routine, short-term testing techniques.

The idea of using large transparent containers in the natural environment is not new. Probably the most well-known examples are the experiments of

Strickland and his associates (Strickland & Terhune, 1961; McAllister, Parsons, Stephens & Strickland, 1961) who studied primary production in a large plastic sphere at the sea surface. Currently, Dr John H. Steele and his co-workers of the Marine Laboratory, Aberdeen are conducting similar experiments in the marine Loch Ewe on the north-west coast of Scotland, and are actively co-operating with CEPEX scientists.

During 1974, the first full year of the CEPEX programme, a series of CEEs (2.5 m in diameter, 15 m deep, 66 m^3 in volume) were tested for their engineering and biological feasibility over periods of 30 days in Saanich Inlet, Vancouver Island, British Columbia, a well-protected, unpolluted fiord known to be biologically highly productive. An initial design limitation of these CEEs was a tendency for limited surface water exchange as a result of wave action during occasional periods of high wind activity. Subsequently, it is planned to use much larger CEEs, up to 2000 m^3 in volume, in order to permit withdrawal of sufficient amounts of biomass at each trophic level for chemical analysis of pollutants and experimentation over periods of 2 or more months.

In some of the first experiments with CEEs, Takahashi *et al.* (1975) described biological changes in four unpolluted CEEs for up to 1 month. They demonstrated that such changes occurred in all the CEEs and ranged over several orders of magnitude with time, but that between CEEs the degree of replication generally differed by a factor of less than two. Such experiments imply that any serious lack of replication between CEEs can be regarded as being due to non-natural effects. The first group of pollutants selected for testing following these replication studies was heavy metals, starting with copper and mercury. Because of the time scale and unit size involved it was obvious that very few ecosystem experiments could be run in any one year, even on the reduced scale of the first year. Short-term laboratory experiments therefore were initiated to ascertain the useful range of types and concentrations of pollutants to be added in the CEEs and to develop and evaluate measurement and monitoring techniques.

Virtually all synthesised organics of ocean ecosystems are channelled through the zooplankton prior to utilisation by the higher and economically important trophic levels. This zooplanktonic assemblage is much more complex than a single herbivore trophic level (Mullin, 1969; Steele, 1974), containing, in addition to herbivores, primary and secondary carnivores, multivores and degraders. The relative stability of this conglomerate to perturbation may well hold the key to the sensitivity of the ecosystem as a whole to such influences. There are still only few data for marine zooplankton on the effects of any pollutants, including copper. In the comprehensive review by Lewis & Whitfield (1974) on the biological importance of copper in the sea, virtually the only recent data for effects on marine planktonic invertebrates come from work on larvae of benthic adults (Bernard & Lane, 1961; Wisely & Blick, 1967; Connor, 1972). The present paper reports progress of the first year of the CEPEX programme with particular regard to results obtained with the larger zooplankton (defined as those caught in nets of 200 μm mesh aperture). It describes both the short-term laboratory experiments and the problems, experience and conclusions to date of the ecosystem experiments.

Short-term acute toxicity laboratory measurements

Methods

Zooplankton organisms for toxicity tests were obtained from the subtropical waters in Biscayne Bay and the Gulf Stream off Miami, and the boreal waters of Saanich Inlet, British Columbia.

For each toxicity test duplicate tests were run for the control and for four to six concentrations of the toxicant. Covered, 190-mm diameter crystallising dishes containing 1700 ml of seawater were used as experimental containers except for very small animals (rotifers, nauplii) and large animals (ctenophores) where 250- and 4000-ml beakers, respectively, were used. Mercury (as $HgCl_2$) and copper (as $CuSO_4$) were used as the toxic materials. Animal populations were counted directly at the beginning and end of each 24-hour experiment. Numbers in each container ranged from 100 of the smaller species (e.g. *Acartia tonsa*) to ten of the larger (e.g. *Sagitta hispida* adults). The experimental medium for Miami organisms was offshore Gulf Stream seawater at 21 °C and 36‰ salinity; for Saanich organisms it was 13 °C and 27‰ salinity.

Zooplankton tested ranged over four phyla, over one entire life cycle (*Sagitta hispida*), between inshore and offshore species of the same group, and over a carbon dry weight range of 0.1–2500 μg. A total of twelve species was involved. Death was usually easy to recognise, but where animals were moribund (incapable of making the normal locomotory movements) and, therefore, not planktonic, they were also counted as dead. This fraction rarely amounted to more than a small percentage of those actually dead. Overall survival of control individuals was 87 %.

Results and discussion

All results are expressed in terms of 24-hour LC_{50} values, i.e. the concentrations of pollutant estimated from inspection of the derived mortality curve which killed 50 % of the organisms in 24 hours. Fig. 1(*a*) illustrates the effect of copper and mercury on different life-history stages of *Sagitta hispida*.

Table 1. *24-hour LC_{50} concentrations for copper and mercury*

Organism	Biomass (μg C)	LC_{50} copper (ppb)	LC_{50} mercury (ppb)
Calanus plumchrus	204	2778	—
Metridia pacifica	8	176	—
Euphausia pacifica	2957	14	—
	273	30	
Pleurobrachia pileus	307	33	—
Phialidium sp.	213	36	—
Copepod nauplii	—	90	—
Larval annelids	—	89	—
Acartia tonsa			
adult	2.4	104–311	34
nauplii	0.028	—	3
Labidocera scotti	39.2	132	—
	47.2	—	31
Euchaeta marina	48.8	188	50
Undinula vulgaris	48.8	192	—
Artemia salina	7.2	—	180
	1.4	—	42
	0.86	2050	27
	0.76	2554	35
Brachionus plicatilis	0.052	100	65
Mnemiopsis mccradyi	2480	29	—
	228	—	15
	185	17	—
Sagitta hispida	146	460	118
	68	394	94
	36	315	79
	23	—	66
	1.4	—	20
	0.2	43	11

Fig. 1(*b*) shows the effect of copper on all organisms tested as a function of their size and the data indicate that three response categories can be distinguished. Fig. 2(*a*) indicates the effect of copper on *Acartia tonsa* at different temperatures in measurements made over a period of 6 months. Table 1 lists 24-hour LC_{50} concentrations, size (in organic content) and, where more than one determination was made, the range of values obtained.

These data enable certain general conclusions to be drawn concerning the relative effects of copper and mercury toxicity:

(1) Toxicity is directly related to the size of the organism. This holds whether different size ranges of the same species are considered, as in Fig. 1(*a*), or whether all the species covering a very wide size range are taken together (Fig. 1*b*). The only exception to this general statement was *Euphausia pacifica*,

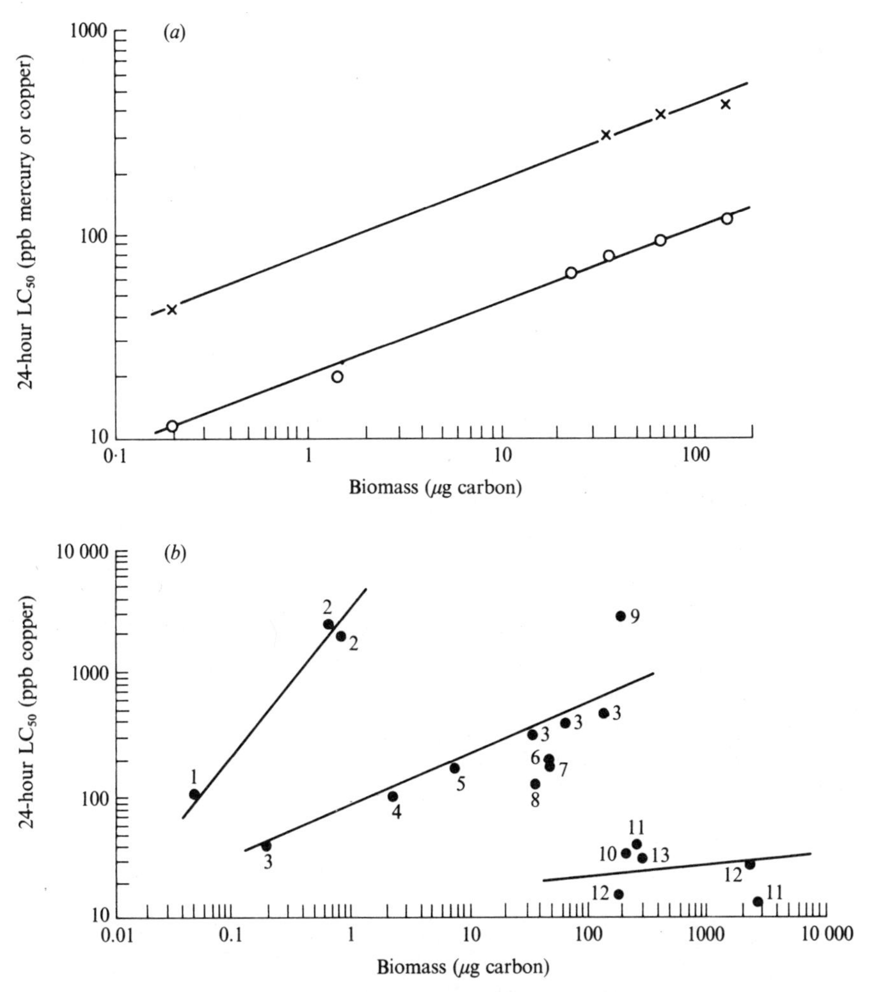

Fig. 1. 24-hour LC_{50} values as a function of biomass in terms of organic carbon. (*a*) values for mercury (○) and copper (×) on a range of sizes of the chaetognath *Sagitta hispida*; and (*b*) values for copper on thirteen planktonic animals. Key to (*b*): 1, *Brachionus plicatilis*; 2, *Artemia salina*; 3, *Sagitta hispida*; 4, *Acartia tonsa*; 5, *Metridia pacifica*; 6, *Undinula vulgaris*; 7, *Euchaeta marina*; 8, *Labidocera scotti*; 9, *Calanus plumchrus*; 10, *Pleurobrachia pileus*; 11, *Euphausia pacifica*; 12, *Mnemiopsis mccradyi*; 13, *Phialidium* sp.

where larger animals appeared to be more sensitive when two size groups were tested. *Sagitta hispida* and *Acartia tonsa*, for instance, both common inshore zooplankton of the south Florida region, appear to have approximately the same sensitivities to a given metal for one size of organism. The adult chaetognath (150 μg C) is much less sensitive than the adult copepod

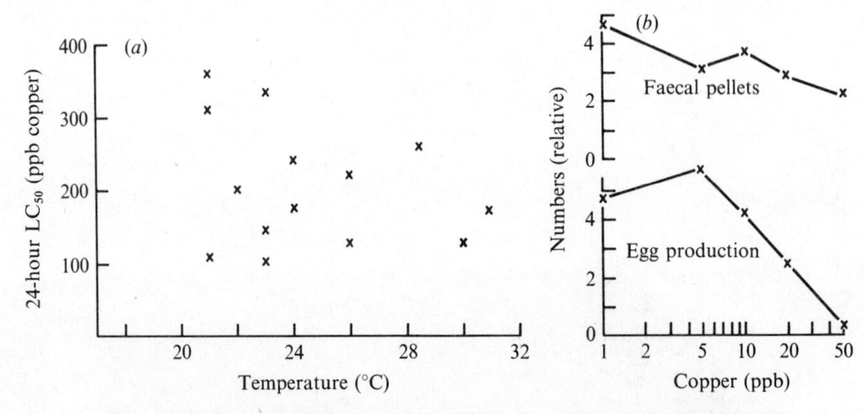

Fig. 2. (*a*) 24-hour LC_{50} for populations of *Acartia tonsa* at various ambient Biscayne Bay water temperatures (i.e. different times of the year). (*b*) Faecal pellet and egg production rate of *Acartia tonsa* exposed to various concentrations of copper over 4 days in the laboratory (numbers are relative).

(2.5 μg C) but this only reflects their size difference. This suggests that sensitivities for intermediate size ranges can, therefore, be predicted when values for larvae and adults have been established.

(2) Species may be grouped into high, average and low sensitivity categories relative to their size. The chaetognath and all the copepods fall within the average group. The two ctenophores (*Mnemiopsis* and *Pleurobrachia*) and the euphausid are much more sensitive for their size, whilst the typical, hardy laboratory animals (*Artemia* and the euryhaline rotifer *Brachionus*) are much less sensitive.

(3) Within a region, there is little difference between the sensitivities of inshore and offshore or oceanic species. Off the Florida coast *Undinula vulgaris* and *Euchaeta marina* are the two commonest large epipelagic species in the Florida Current, and *Labidocera scotti* is a neritic copepod intermediate in its ability to withstand changes in temperature and salinity. All three are a little more sensitive to both mercury and copper than the predicted values based on the line drawn for the inshore/estuarine *Acartia tonsa* and *Sagitta hispida*, but these differences are small compared to the effects noted in (1) and (2) above.

(4) Temperature also plays a minor role in affecting sensitivity. The copepods *Metridia pacifica* and *Calanus plumchrus* from Saanich Inlet fall close to the 'average' line. The Saanich 'jellies', *Pleurobrachia* and the medusa *Phialidium*, are only slightly less sensitive than the warm water *Mnemiopsis* (Fig. 1*b*). LC_{50} determinations made on *Acartia tonsa* over 5 months are so

scattered that no conclusions can be drawn concerning changes in sensitivity with respect to temperature (Fig. 2*a*).

(5) Populations of the species *Acartia tonsa* vary considerably in sensitivity to copper, as indicated in Fig. 2*a*. Over a period of 6 months 24-hour LC_{50} measurements ranged from 100 to 300 ppb copper, even at similar ambient environmental temperatures. There appears to have been a variation in response to copper of different populations separated temporally and in terms of generations. In a separate series of six tests made over an interval of 10 days, a range of only 120 to 170 ppb was obtained. This implies that there may be wide genetic variability within a species in its tolerance of toxic substances, and that populations which might initially be decimated in a particular locality may subsequently return to their previous population density over a longer period.

(6) Organisms are some two to four times more sensitive to mercury than to copper.

In summary, short-term acute toxicity measurements can be useful in providing the maximum amount of comparative data for the shortest investment of time. They serve well to screen different compounds within a class (e.g. the fractions of petroleum crude oil), to establish relative toxicities prior to more detailed studies of the most toxic components or to screen species within a community to determine which are the most sensitive, relative to a particular compound.

Sub-lethal toxicity laboratory measurements

The purpose of these experiments was to obtain a better estimate of suitable dosing levels for the chosen pollutant in the CEEs, and to develop a sensitive, standard testing procedure which could predict accurately the long-term effects observed in the CEEs.

Originally it appeared that attempts should be made to culture one or two species throughout their life cycle under toxic stress. However, since the data reported above demonstrated that the earliest stages of the life-history are the most sensitive, it appeared expedient simply to expose the egg production and hatching processes to the toxic stressor.

Methods

Adults of the copepod *Acartia tonsa* were maintained in large (60-l), tall (400-mm) aquaria over 4 days at densities of 50-l^{-1}, and fed mixed cultures of the algae *Phaeodactylum* and *Dunaliella* replenished daily to an approximate level of 5 mg chlorophyll *a* m^{-3}. Each day, three 250-ml beakers were placed as sedimentation traps at random on the bottom of every aquarium for a period of 16 hours overnight (during which time the aquaria were in total darkness and not aerated) to permit uniform distribution of copepods and settlement of eggs, faecal pellets and dead animals. In the morning the sedimentation traps were removed, their contents combined for each aquarium, and aliquots counted. Using a population of the copepod *Acartia tonsa* for which the 24-hour LC_{50} had been determined to be about 300 ppb Cu, five aquaria were set up at 21 °C with initial copper levels of 0, 5, 10, 20 and 50 ppb Cu. No attempt was made to maintain these copper levels. Cell cultures were grown without chelators.

Results and discussion

Fig. 2(*b*) summarises some of the results of this experiment. Mortality of the adult copepods over the 4 days was variable, ranging from 16–36 % with a mean of 25 %. There was no consistent trend with copper concentration, indicating no gross mortality effects in the higher copper levels. A clear trend downward in feeding activity and egg production occurred between 10 and 20 ppb. At 50 ppb, faecal pellets were reduced to half the control level, while egg production was almost eliminated. At a temperature of 21 °C the effects were visible even within 24 hours of exposure, indicating that these two parameters could be used to demonstrate sensitivity at copper concentrations 1–1.5 orders of magnitude lower than the 24-hour LC_{50} concentrations (approximately 300 ppb) measured for the same population.

Populations in the controlled ecosystems

Methods

Two separate experiments were conducted in 1974 using copper as the toxic material. The first, lasting 25 days starting on 15 June, consisted of four CEEs; two with no copper (control), one with 10 and one with 50 ppb Cu. The second, starting on 10 September and lasting 20 days, consisted of three containers; one control, one with 5 and one with 10 ppb Cu. The water temperature throughout the column ranged between 11 and 17 °C in both June and September.

Zooplankton samples were collected from the CEE with a 200 mm modified Bongo frame equipped with a 202 μm mesh aperture. Each tow sampled approximately 0.44 m^3 of water. The sampling schedule consisted of three replicate tows from 14 m to the surface in each container every 4 days.

Sampling was initiated 1 day before the introduction of copper into the containers.

From each sample macroscopic 'jellies' (ctenophores and medusae) were removed, identified, measured and counted. The sample was then divided by means of a plankton splitter. An aliquot containing approximately 200 animals was counted under a dissecting microscope. For samples collected early in the experiment an aliquot of $\frac{1}{16}$ or $\frac{1}{32}$ of the original volume was usually examined, but as the experiment progressed and the total number of organisms in the CEEs declined it was necessary to enlarge the subsample, until in some cases (particularly the copper-containing populations) 100 % of the sample was counted. The jellies were counted from the entire sample.

Results and discussion

Probably the most striking feature of both experiments is the sharp decline in the total zooplankton that occurred in all containers during the experiment. Decline was sharper in the copper-containing CEEs than in the controls, with the one containing the highest copper level (50 ppb Cu) exhibiting the greatest decline.

The following remarks specifically relate to the first (June) copper experiment, but estimates of total numbers from the second experiment confirm these results in general. By day 9 in the June experiment (Fig. 3*a*) the two control populations were reduced to 61 % and 50 % of their initial levels, and the 10 and 50 ppb CEEs were 12 % and 3 %, respectively, of their original numbers. On the final day of sampling (day 25) the controls had 15 % and 5 % and the other CEEs 6 % and 0.5 % of their day 1 zooplankton count. Of the ten most abundant groups seven (*Acartia longiremis, Pseudocalanus minutus*, larval barnacles, echinoderms, gastropods, larvaceans and pelecypods) exhibited the same general pattern of abundance as the total zooplankton, showing decrease in all CEEs with the decline in the 50 ppb CEE significantly greater than that in the control CEEs and the 10 ppb CEE intermediate. The number of jellies (Fig. 3*b*) fluctuated in the controls and even increased in the last days of the experiment due to an increase in the number of small medusae. The number of jellies in the 10 ppb CEE decreased steadily after the addition of the copper and reached zero by day 25. Jelly populations in the 50 ppb CEE declined to 11 % of the initial value immediately after copper addition and reached zero by day 17. Two diversity indices, Shannon-Wiener's '*H*' (Shannon & Weaver, 1964) and 'expected number of species' (Hurlbert, 1971) were computed for each sample. Neither index revealed a pattern of change in diversity which could be related to copper concentration.

Relationships among samples were further investigated by the computation

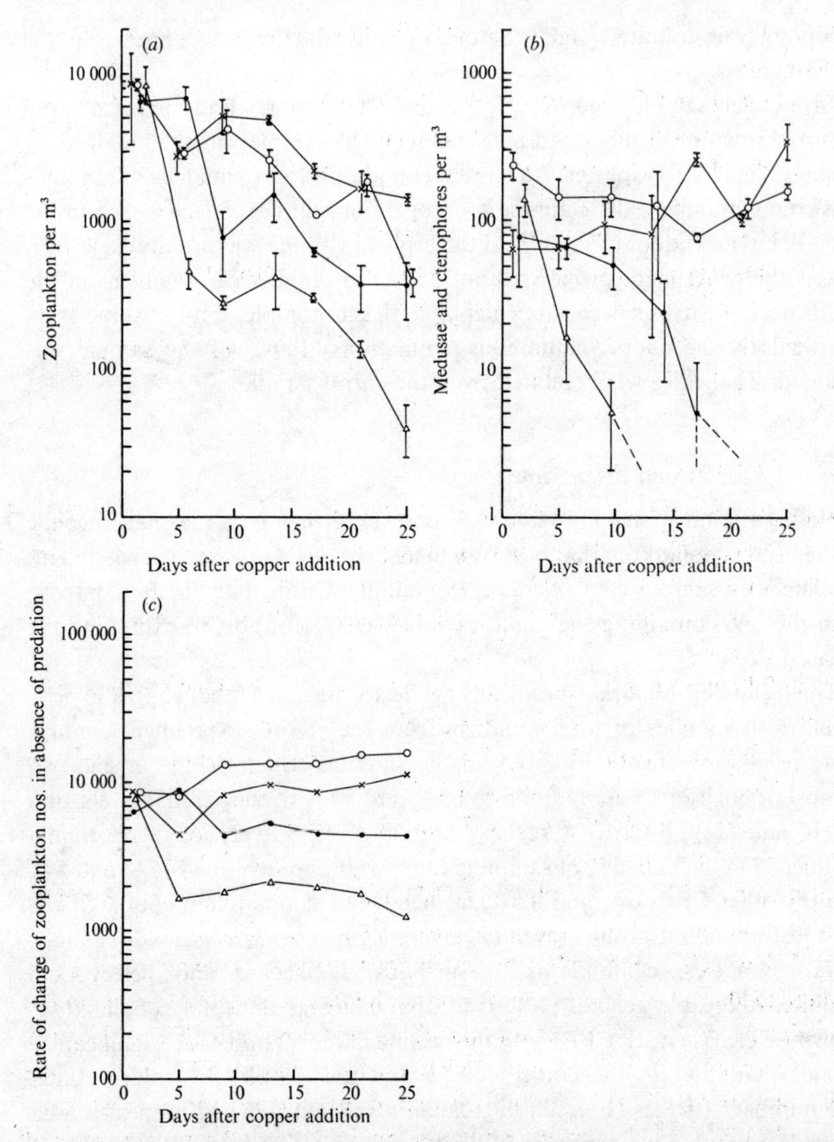

Fig. 3. The rate of change following copper addition in the CEEs of: (*a*) total zooplankton (excluding medusae and ctenophores); (*b*) medusae and ctenophores; and (*c*) the computed rate of change of total zooplankton numbers in the absence of predation. Vertical bars in (*a*) and (*b*) indicate standard deviation of replicates. Open circles and crosses represent control CEEs, closed circles and triangles represent CEEs with 10 and 50 ppb copper respectively.

of similarity indices (Whittaker, 1952) between each pair of samples. These were then clustered by using the unweighted pairs linkage method (Sokal & Sneath, 1963). From the dendrogram (Fig. 4) it is evident that the samples are divided into two groups: control (group I) and polluted (group II). Group I contains all samples from one control (CEE OA) and all but four samples from the other control (CEE OB) as well as all but one sample from the 10 and 50 ppb CEEs from the first two sampling periods (days 1 and 5). Group II contains samples primarily from days 9–25 of CEEs 10 and 50, i.e. the polluted samples. It can also be seen from the dendrogram that replication quality throughout the experiment was quite good. Replicates are usually clustered together at a high level of similarity and where they are not they are at least more closely related to each other than to other samples. Fig. 4 also provides evidence that all CEEs had similar community structure on day 1, since all replicates on Day 1 are clustered together in group III at a similarity level of 0.82.

There are at least three possible causes for the reduction of total zooplankton in all the CEEs. The first might be an inherent effect arising in the bags themselves, either by virtue of toxic materials leaching out from the polyethylene CEE fabric, or from fouling micro-organisms on their surface. This possibility had been checked earlier by J. G. Gamble & Reeve (unpublished data) who found no differences in feeding rates of euphausids in water taken from within and outside the CEEs. The other two obvious causes of mortality are predation by the carnivorous medusae and ctenophores in all the containers, and the effects of copper in those to which it was added.

There can be little doubt that the carnivores were largely responsible for the drastic decrease in the other zooplankton in the control CEEs in both experiments. A similar phenomenon occurred in the Loch Ewe CEEs. Total numbers of carnivores did not decrease with time, and the appearance of very small organisms towards the end of the experiment suggested that reproduction was occurring in the control containers. Because of the overshadowing effects of predation, a clearcut statistical demonstration of mortality effects due to copper was not possible. Copper-related mortality can be inferred from inspection of Fig. 3(*a*) and (*b*) however, because there is an inverse relationship between predator numbers with time and copper concentrations as there is with numbers of other zooplankton, i.e. although zooplankton mortality was higher with increasing copper concentration the predation pressure was progressively lower.

An attempt was made to estimate the predatory effect of the carnivorous jellies, in order to obtain a basis on which to adjust values obtained for the standing stock of other zooplankton in the CEEs over the first few days and so to separate the fraction of mortality due to the copper effect. To accomplish this adjustment, measurements were made of the respiratory rate of various

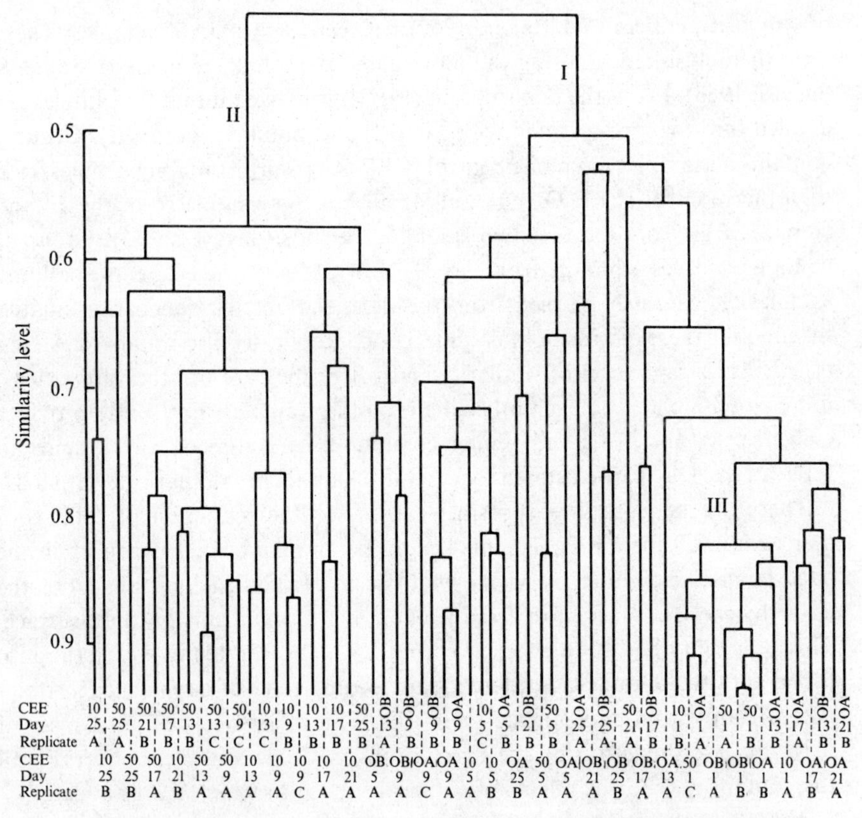

Fig. 4. Similarity dendrogram. Samples from control (OA and OB), 10 and 50 ppb Cu CEEs clustered by the unweighted pairs linkage method. Values are for replicates A, B, C and days 1, 5, 9, 13, 17, 21 and 25.

sizes of the ctenophore *Pleurobrachia* and the medusa *Phialidium* and a relationship was established between body diameter and respiration rate. Taking a respiratory quotient of 0.8 and a carbon content of dry food material of 50 %, the dry weight of food required to account for respiration was calculated. This dry weight was then related to plankton depletion after assuming that the food intake required by the predators for growth (allowing for some incomplete assimilation), would be twice the respiratory requirement. In Fig. 3(*c*), the numbers of other zooplankton have been adjusted on this basis from the observed to a theoretical number assuming no predation. Conversion to numbers from dry weight values was made by using the average dry weight of 5.8 μg per food organism from a control CEE. Over the first 4 days of the experiment the jellies were calculated to have removed 3570, 2130 and 1220 food organisms per m^3 in the 0, 10, and 50 ppb CEEs, respectively, and during the first 8 days 6140, 3530 and 1550 organisms per m^3.

At the end of the experiment the percentages of the original numbers of total zooplankton remaining were 10, 6 and 0.5 % for, respectively, the controls (averaged), 10 and 50 ppb containers (Fig. 3*a*). Corrected for the computed predation effects, these values would have been 149, 64 and 19 % (Fig. 3*c*). It should be emphasised that these calculations are very approximate and become progressively more unreliable beyond the first few days. Ingestion rate in ctenophores is a function of food concentration and feeding tends to be superfluous at high concentrations (Baker & Reeve, 1974), which would tend to result in underestimation of their predatory effect at the beginning of the experiment. On the other hand, respiration rates were calculated for healthy animals not exposed to copper, but the fact that jelly numbers declined drastically at 10 and 50 ppb Cu suggests that many of the surviving jellies in those containers were probably not healthy and therefore may have consumed less than the healthy animals.

As noted earlier, population changes in the second (September) copper experiment, at 0, 5 and 10 ppb, bore a remarkable similarity to those at 0, 10 and 50 ppb in June. Total zooplankton excluding ctenophores decreased to 18, 6 and 0.5 % and jellies to 39, 6 and 0.1 % of their original numbers at 0, 5, and 10 ppb Cu respectively.

Sub-lethal toxicity effects on experimental ecosystem zooplankton

In the first copper experiment in June 1974, participants agreed that a withdrawal rate of zooplankton of no more than 1 % per day should be made from the CEEs. Since population analysis required sampling close to the maximum rate, organisms for experimental purposes were not taken from the CEEs. Instead, plankton collected from the environment adjacent to the containers were used in experiments.

Methods, results and discussion

Effect of copper on feeding rate. It is well established (see Parsons & Takahashi, 1973) that feeding rate of herbivorous zooplankton is a function of food availability, both in terms of absolute concentration and sizes suitable for effective capture by the filtering appendages. In both copper experiments phytoplankton populations in the treated CEEs tended to be dominated by microflagellates and diatoms less than 10 μm in diameter, while the untreated CEEs were dominated by large *Chaetoceros* chains. Additionally, larger absolute concentrations of chlorophyll *a* developed during both experiments

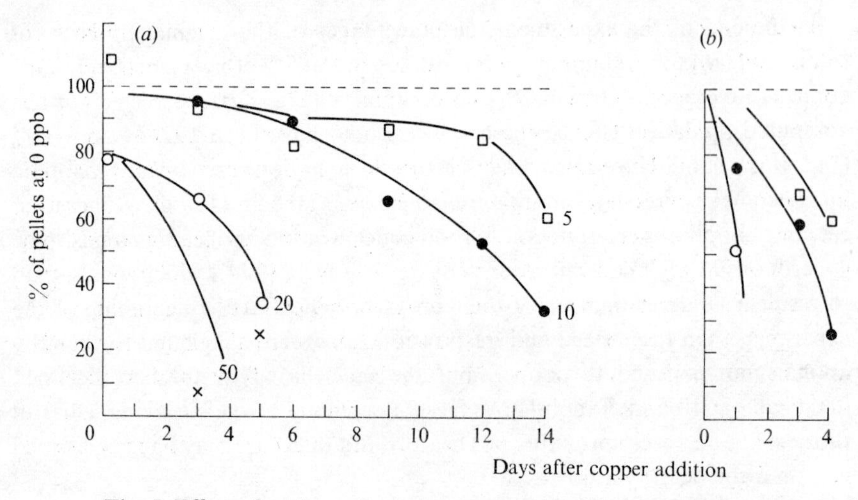

Fig. 5. Effect of various copper concentrations (ppb) on faecal pellet production of (*a*) *Calanus plumchrus* and (*b*) *Metridia pacifica* in the laboratory in relation to pellet numbers produced by uncontaminated copepods (horizontal dashed line).

in the copper-containing ecosystems, probably because the predation effects of the herbivorous zooplankton decreased more rapidly. An experiment which compared feeding (in terms of either filtering or ingestion rates) of copepods from each CEE on the phytoplankton population of its own CEE would therefore be expected to yield different values, independent of any effect due to copper. This problem can be overcome by feeding animals on a standard source of food, such as the phytoplankton population in the waters outside the containers at the time of the experiment.

During the June copper experiment the copepods *Calanus plumchrus* and *Metridia pacifica* from outside waters were maintained at copper levels up to 50 ppb. Groups of these animals were placed in 1-l jars in water at the same copper concentration from which they came, and which contained food (diatoms) at the same concentration in all jars. These vessels were rotated slowly (1 rev. per 2 min) on a wheel overnight. The following morning the resulting faecal pellets were allowed to settle and were collected and counted. The results are presented in Fig. 5 in terms of the number of pellets produced at any copper concentration as a percentage of those produced at 0 ppb by the same species on that day. The results clearly show the effect of copper at sublethal concentrations on faecal pellet production and, by inference, its effect on the feeding process. *Calanus*, which had a 24-hour LC_{50} of over 2000 ppb Cu, was seen to be increasingly affected over a period of 14 days even at a copper concentration of 5 ppb. The effect occurred more rapidly as copper concentration increased. The smaller *Metridia*, with a 24-hour LC_{50} of

200 ppb, produced reduced numbers of pellets within 4 days of exposure to 5 ppb.

This effect was subsequently confirmed in a lengthy series of tests during the September copper experiment when ingestion rates were determined by measuring decrease in food cell density over the period using a Coulter counter, and test animals were withdrawn directly from the CEEs.

Effect of copper on respiration and excretion. As with feeding, these parameters were also measured both in populations exposed to copper in the laboratory and those withdrawn directly from the containers. Preliminary experiments were run to test whether there were any measurable changes in oxygen consumption and ammonia production in animals within the first few hours of exposure to copper and mercury. Significant changes in these parameters did not occur until concentrations were reached which were close to the lethal dosage; measurements of respiration and ammonia production are therefore not particularly useful as more sensitive indicators of stress over the short-term than LC_{50} values.

Longer-term tests on laboratory-maintained animals produced similar conclusions. *Calanus plumchrus* populations were maintained in the laboratory at 13 °C for up to 15 days in 10 ppb Cu and 25 days in 5 ppb Cu. There was no indication after 25 days that 5 ppb reduced the respiration below that of the unexposed animals, despite the fact that the population at 5 ppb had finally died out by that time, suggesting that they were subjected to greater stress than the control animals. In the medusa *Phialidium* and the ctenophore *Pleurobrachia* there were obvious reductions in respiration at these copper concentrations, but both species were close to their lethal levels at such concentrations (see Table 1) as evidenced by abnormal swimming behaviour.

It seems likely that respiration and ammonia excretion have not proved very useful as sensitive indicators of stress because there are many other factors such as feeding condition and laboratory handling and maintenance (T. Ikeda, unpublished data), which also influence the underlying metabolic processes.

Discussion

The purpose of this paper has been to present an account of the approaches which have been used from the beginning of the CEPEX programme up to the end of the first (June, 1974) CEE copper experiment to evaluate the chronic effects of pollutants in the marine environment, specifically with regard to the larger zooplankton. Since the full-scale CEEs have not yet been used, all the work to date has been exploratory in an effort to minimise potentially costly omissions or errors in the management of, and scientific measurements made in connection with, the 2000 m^3 CEEs.

During both copper experiments the dominant phytoplankton in the waters surrounding the CEE were large chain diatoms. A common phenomenon within a few days of capture of the water column of the 66 m^3 CEEs, in both the June and September experiments, in the earlier replication studies (Takahashi *et al.*, 1975) and in the Scottish experiments, was a 'crash' or rapid population decline of the phytoplankton. Whether this was a natural event unconnected with the capture of the water column is not known. It raised, however, the question as to whether a similar 'capture effect' could be a contributing factor in the decline of zooplankton populations. In this connection, the Scottish group ran a similar 1-month 10 ppb copper experiment in Loch Ewe, and found no clear mortality effects obviously attributable to copper. The Scottish results were similar to those of CEPEX in that large numbers of ctenophores produced severe mortalities in both control and copper CEEs which tended to obscure other mortality effects. The Loch Ewe experiments differed in that the water column had been captured 1 month prior to the addition of copper. J. H. Steele (personal communication) suggested that their populations might have become adjusted to CEE conditions and therefore less susceptible to the added stress of copper. In the second CEPEX copper experiment the populations were captured 11 days prior to copper addition, by which time the phytoplankton standing crop had reached low levels in all the CEEs. Since this experiment produced clear evidence of mortality in excess of predation (and predators themselves were progressively reduced in the copper-containing CEEs), a factor related to the copper additions was clearly operating. The experimental work reported above lends support to the assumption that it was a simple copper effect, but the synergistic effect of confinement in the CEE itself cannot be discounted. The much larger volume and longer experimental time period in the full-size CEEs should greatly reduce the possibility of any container effect.

A second suggestion by Steele was that the Scottish copepod populations were more resistant to copper because the ambient ionic copper content of Loch Ewe water was 2 ppb compared to 0.2 ppb in Saanich Inlet. The possibility of distinctly different copper tolerances exhibited by closely related organisms of the same size appears convincing from the data of Fig. 2(*a*), where wide differences are recorded from populations of the same species separated by time but from the same location and temperature. We intend to examine further the capability of populations to acquire resistance to pollutants. A third possibility, which has not yet been directly tested. is that the natural copper-chelating or complexing capacity of the Saanich Inlet and Loch Ewe waters is different.

One of the main problems in the CEEs is replication. The experimental programme in the CEEs is based on the concept of a series of controlled experiments. It has been demonstrated fairly clearly that, at least over 1 month, replicate results can be obtained from different CEEs (Takahashi *et al.*, 1975) provided that they contain similar populations initially. The introduction of pollutants, however, produces differential effects which would normally produce differential responses independent of any copper addition. These have been referred to above and include gross effects, such as differential predation pressure, and sublethal effects, such as variations in the amount and kind of available food. We do not know, as yet, whether the full-scale containers can be filled so as to capture similar populations, nor whether control populations will replicate and maintain species characteristic of the local waters for periods much in excess of 1 month.

The initial CEE experiments have demonstrated that, at least in the short- and intermediate-term (up to 1 month), the zooplankton are probably the most affected component of the system. This is not because they are inherently more sensitive to copper as a group, but, being higher up the food chain, there are fewer species and numbers and they have longer life cycles. It would take from 1 to several months (or years in the case of the larger species) for resistant members of one species, or resistant species, to repopulate the environment at former biomass levels if 90 % or more of the total zooplankton population had been eliminated. It appears from the work of our colleagues that copper-resistant species of heterotrophic micro-organisms can develop large populations within a very few days, even at 50 ppb Cu, by virtue of their extremely short generation time. In the case of the phytoplankton some common species were eliminated but others replaced their biomass within 10–20 days.

Conventional short-term LC_{50} laboratory experiments yield rapid data on comparative effects of pollutants and can be very useful in screening hundreds of potentially toxic substances, but the results have little to do with assigning 'safe' or permissible levels of such toxicants in the oceans. The results of the CEPEX programme to date are beginning to demonstrate the problems and pitfalls of extrapolating simple laboratory data to natural conditions.

The authors wish to record their grateful thanks to Dr M. Takahashi, chief scientist at the CEPEX site, as well as Mr P. Koeller and F. Whitney and all the other support staff and students. This work was made possible by the support of NSF (IDOE) grants GX-39140 and GX-42580.

References

Baker, L. D. & Reeve, M. R. (1974). Laboratory culture of the lobate ctenophore *Mnemiopsis mccradyi* with notes on feeding and fecundity. *Marine Biology*, **26**, 57–62.

Bernard, F. J. & Lane, C. E. (1961). Absorption and excretion of copper ion during settlement and metamorphosis of the barnacle, *Balanus amphitrite niveus*. *Biological Bulletin*, **121**, 448–50.

Connor, P. M. (1972). Acute toxicity of heavy metals to some marine larvae. *Marine Pollution Bulletin*, **3**, 190–2.

Hurlbert, S. H. (1971). The non concept of species diversity: a critique and alternative parameters. *Ecology*, **54**, 577–86.

Lewis, A. G. & Whitfield, P. H. (1974). The biological importance of copper in the sea, a literature review. *International Copper Research Association Report no. 223.*

McAllister, C. D., Parsons, T. R., Stephens, K. & Strickland, J. D. H. (1961). Measurements of primary production in coastal sea water using a large-volume plastic sphere. *Limnology and Oceanography*, **6**, 237–58.

Mullin, M. M. (1969). Production of zooplankton in the ocean; the present status and problems. *Oceanography and Marine Biology Annual Review*, **7**, 293–314.

Parsons, T. R. & Takahashi, M. (1973). *Biological oceanographic processes.* Pergamon Press, Oxford. 186 pp.

Shannon, C. E. & Weaver, W. (1964). *The mathematical theory of communication.* University of Illinois Press, Urbana, Illinois. 125 pp.

Smith, W. & Grassle, J. F. (1975). A diversity index and its sampling properties. *Ecology*, in press.

Sokal, R. & Sneath, P. H. A. (1963). *Principles of numerical taxonomy.* W. H. Freeman and Co., San Francisco. 359 pp.

Steele, J. H. (1974). *The structure of marine ecosystems.* Harvard University Press, Cambridge, Mass. 128 pp.

Strickland, J. D. H. & Terhune, C. D. B. (1961). The study of *in situ* marine photosynthesis using a large plastic bag. *Limnology and Oceanography*, **6**, 93–6.

Takahashi, M., Thomas, W. H., Seibert, D. L. R., Beers, J., Koeller, P. & Parsons, T. R. (1975). The replication of biological events in enclosed water columns. *Archives für Hydrobiologie*, in press.

Whittaker, R. H. (1952). A study of summer foliage insect communities in the Great Smokey Mountains. *Ecological Monographs*, **22**, 1–44.

Wisely, B. & Blick, R. A. P. (1967). Mortality of marine invertebrate larvae in mercury, copper and zinc solutions. *Australian Journal of Marine and Freshwater Research*, **18**, 63–72.

G.M.HUGHES

Polluted fish respiratory physiology

Research is a very varied field of human endeavour and those involved in it are motivated for different reasons. Some prefer a specific goal with a definite application; others are more attracted by research into problems for their own sake. Sometimes such problems arise from teaching or in relation to specific human problems such as disease. Sometimes they arise from other researches, or they are problems dreamt up by the individual scientist concerned. What is most important to both of these main sets of workers is that they investigate problems about which they are enthusiastic and are keen to solve. The separation of these two types of approach into 'applied' and 'pure' science, or whatever other names they are given, sometimes misses this important fact. It also tends to suggest that neither type of scientist is interested in the other's type of problem. Furthermore it must not be forgotten that both types of scientist can make fundamental scientific discoveries. It would seem best, from our knowledge of the history of science, to accept the existence of such differences between scientists and to recognise that they both have important roles to play.

One feature of pollution studies which I find attractive is the great variety of problems involved and the varied backgrounds of the investigators. Multidisciplinary approaches are generally very fruitful but they require a special type of tolerance and understanding of other viewpoints. As is generally true of such fields there are many workers in the main stream of the research but others can act as occasional consultants or collaborators who hope to introduce new ideas and techniques with which they are especially familiar. Often workers in pollution call upon such people and the integration between scientists in specialised institutes and those involved in academic research is, in my own opinion, to be encouraged.

In pollution research, the experimental biologist finds many interesting problems and presumably one of the final goals is to reduce the levels of pollution to those which can at least be tolerated and, dare I say, be beneficial to the fish and other populations. Some scientists are immediately attracted by problems of this kind but equally good scientists are completely uninterested. Normally, because of their broad plan of work, fish physiologists will be concerned with problems which have arisen in teaching and which are completely separate from any applied connotation. Inevitably, however, and because they are dealing with fish and with interests in their environment, some will become involved in problems which have a bearing on pollution. Equally inevitably, the applied worker will find the need for information about the 'norm' of the fish which can often only be provided by the other type of worker.

It may appear that some of these generalisations are theoretical but in fact they are based directly on my own experience in this field. In 1959, I received a letter from Dr Herbert of the Stevenage Water Pollution Laboratory asking if they could visit Cambridge in order to show me some sections of gills of fish from waters containing suspended solids (Herbert & Merkens, 1961) and discuss some problems which had arisen during studies of Mr Lloyd on the effect of heavy metal pollution on fish. During their visit, I was able to show them some of our work and also to tell them about the current state of the literature; for example I showed them the notable thesis by van Dam (1938). In this thesis, and our own work (Hughes & Shelton, 1958), measurements were made of the ventilation volume of fish, i.e. the volume of water pumped through the gills in unit time. Of course there had been earlier studies making use of physiological techniques in fish pollution research but these had only been concerned with recording the frequency of the ventilatory movements. From the point of view of pollution it is the volume of water passing through the gills which is so much more significant. Lloyd (1961) showed, for example, that the effect of different pollutants was greater at lower oxygen tensions when the ventilation volume is increased. The gills clearly come into close contact with this water and hence are readily affected by the pollution.

Fig. 1 summarises diagrammatically the whole system involved in respiration and circulation of a fish. Water and blood are pumped through the gill exchanger and both the ventilatory and cardiac pumps can vary their total outputs by changes in either frequency or stroke volume. The diagram also emphasises how both pumps consume oxygen and it is because of this that the oxygen supply to the fish under poor conditions of supply or interference with the gills can become limiting. At the fiftieth anniversary meeting of the Society for Experimental Biology in Cambridge (Hughes, 1976) I played a recording made using a Doppler flowmeter which showed large variations in water velocity within the respiratory system of trout. Such fluctuations may partly account for differences between theoretical calculations and measurements of the oxygen cost of breathing. Interference with this flow by pollutants would further increase the oxygen cost of ventilation.

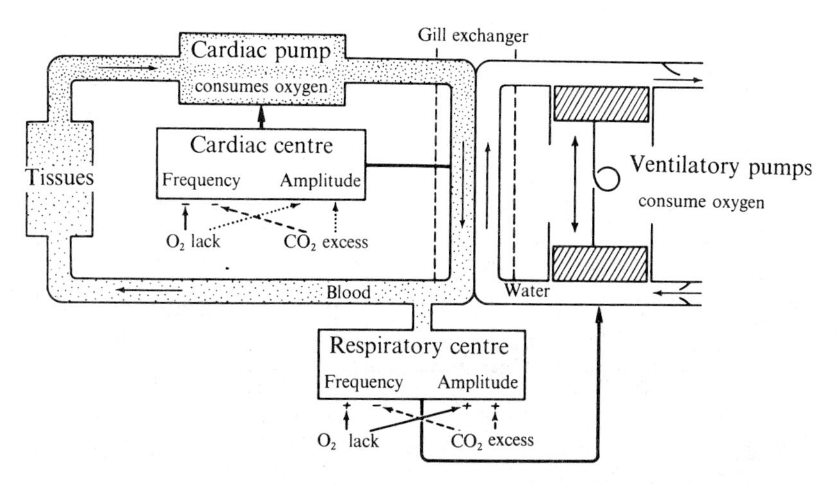

Fig. 1. Conceptual diagram of systems involved in fish respiratory homeostasis. Nerve pathways involved in feedback control of the cardiac and respiratory pumps are indicated by solid arrows for oxygen and dashed arrows for carbon dioxide. Excitatory effects are indicated by +, and inhibitory effects by −. (After Hughes, 1964.)

Thus I was well aware of the pollution problem more than fifteen years ago and the possibility that our particular methods and background knowledge might help in such investigations. Consequently one of our first studies at Bristol showed that suspended solid pollutants such as wood pulp fibre increase the coughing frequency of rainbow trout (Fig. 3). The recording technique used was that developed for measuring pressure changes in the respiratory cavities of fish. The fish is maintained in a closed circulation (Fig. 2); a method subsequently used for many studies of fish respiration. It has the advantage of giving information not only about the frequency but also the depth of ventilation. As with any method of this kind it has the disadvantage of interfering with the animal but fish treated in this way show a normal ventilatory rhythm and other respiratory parameters after about 24 hours in the apparatus and survive for months with the buccal cannula intact. This technique was also used by Skidmore (1970) for studies on the effects of zinc pollution on fish respiration.

That the respiratory system of fish is affected by heavy metal pollution has been appreciated for a long time. I have not discovered who was the first person to make this suggestion but certainly Paul Bert (1871*a*,*b*) concluded that freshwater fish died in seawater because of an arrest of the branchial circulation induced by gill shrinkage. Other studies have also emphasised the osmotic effect of pollutants and it seems clear that very hypertonic solutions may be fatal for this reason. In isotonic solution different salts vary in the rate

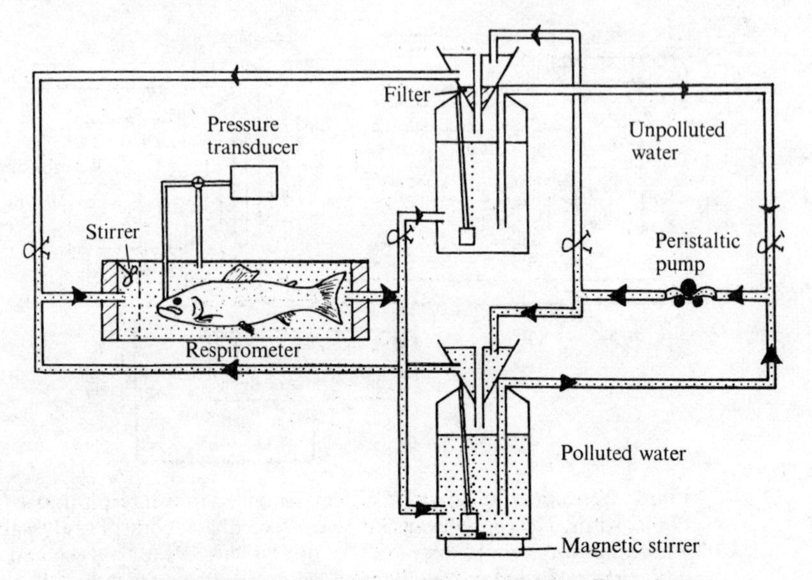

Fig. 2. Diagram of the circulation used for testing the effect of water containing suspended solids. Water flows in the direction of the large arrows and is shown circulating polluted water. It can be changed to circulate unpolluted water by transferring each of the four clips so that flow follows the smaller arrows. (From Hughes, 1975.)

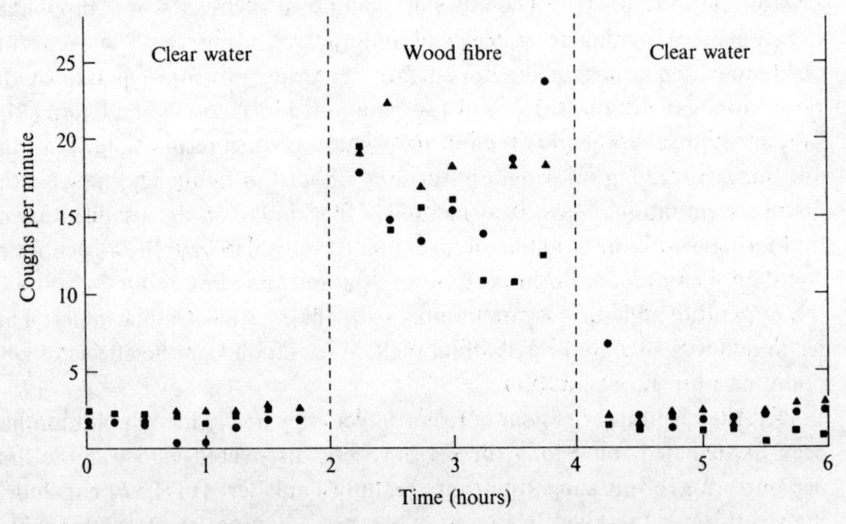

Fig. 3. Plots showing changes in coughing frequency of rainbow trout during three experiments in which the fish were subjected to suspended wood pulp fibre in the circulated water. (From Hughes, 1975.)

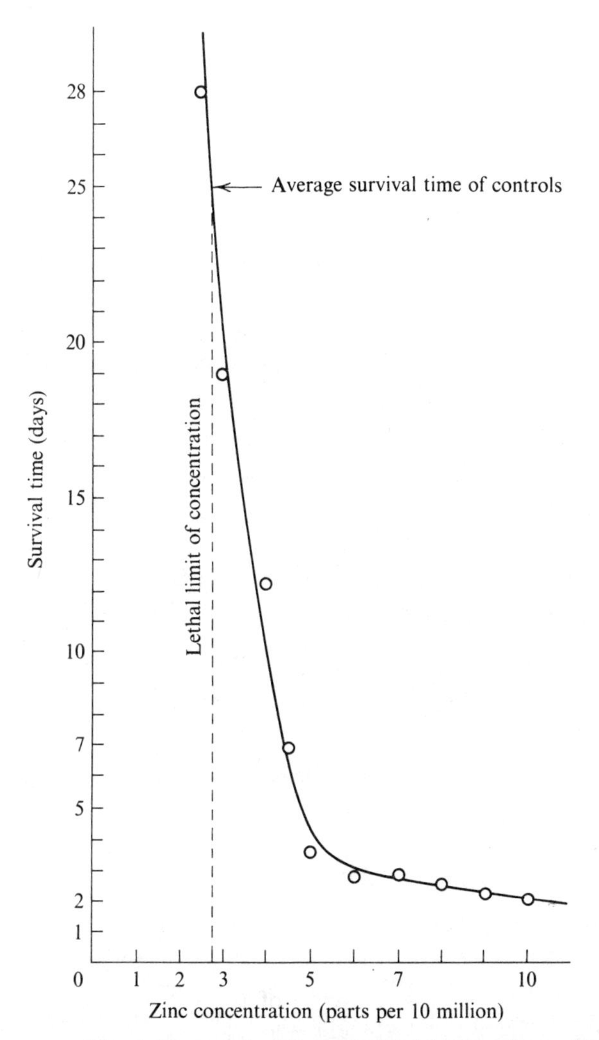

Fig. 4. The survival time of sticklebacks when subjected to different concentrations of zinc. (From Jones, 1938.)

of their action, presumably due to different permeability properties of the gill membranes. There are many examples of the salts of metals being toxic at very low concentrations, well below isotonicity. The exact way in which these salts enter the body and the nature of their toxic action continues to be investigated; there is also evidence that some of them exert their effect without entry to the body, i.e. their effect is primarily on the external surface of the fish and especially the gills.

A typical survival curve for a fish in toxic solutions of a heavy metal is shown in Fig. 4. For laboratory investigations it is clearly desirable to use concentrations that can be maintained throughout the experimental period and which produce a significant effect upon the fish in a relatively short time. Such concentrations have been used by many investigators but some of them are, of course, non-physiological or non-ecological in the sense that fish are rarely subjected to such concentrations even in polluted waters. The survival curve shows a relatively sharp increase in survival time below a certain concentration and it is in this zone that the experimental biologist is best able to apply his techniques. In the case of the stickleback graph this range is 3–6 parts per 10 million of zinc, where the survival time ranges from 2 to 20 days. Ideally studies over longer periods and for pollutant concentrations which are non-lethal would be most valuable in relation to the concentrations found in the environment. In some recent work in collaboration with the Stevenage Laboratories we have been fortunate enough to use specimens that have been treated for long periods as, for example, in the case of cadmium concentrations of 0.002 ppm and fish which were sampled after 7 months' exposure without any deaths. The availability of techniques sufficiently sensitive to detect the effects of such low concentrations is clearly of paramount importance for this type of study. Some of these are discussed below.

Recording of physiological parameters of respiration

Physiological methods directed at monitoring changes in the respiration of fish were carried out many years ago by Jones (1938, 1939, 1964). His contributions in this field are important both in themselves and also because of the stimulation they provide for others who continue this type of work. In some of his work particular attention was given to measuring oxygen uptake and frequencies of the opercular movements (Fig. 5). The observed increase in oxygen uptake is probably due to greater activity of the fish, which includes both general locomotory activity and also that of the ventilatory pumps. Indeed one of the likely effects of pollution affecting the respiratory system is that it limits the metabolic scope for activity, i.e. the difference between the resting and active levels of metabolism. Such an effect would arise because of both an increase in the resting metabolism and a decrease in the upper levels of active metabolism. The technical difficulties encountered during this type of experimentation can only be fully appreciated by those who have attempted such studies, and it is clear that some of Jones' data would have been improved by more recent techniques. For example, the frequency of ventilation in sticklebacks is normally about 100 min^{-1}, which is about the limit that can be observed accurately by eye. Apart from the influence of the observer on the fish, it becomes impossible to make accurate counts when the frequency rises to over 200 min^{-1}. In his description, Erichsen Jones makes this quite clear:

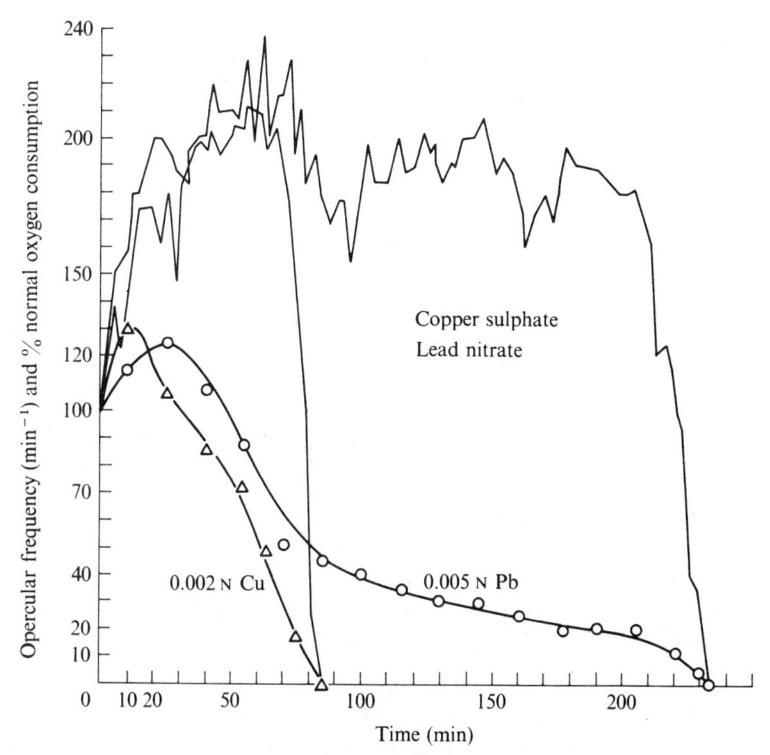

Fig. 5. Oxygen consumption (heavy lines) and opercular movement rate (light lines) for sticklebacks in copper sulphate and lead nitrate solutions at 17 °C. (From Jones, 1964.)

'The fish usually remained sufficiently still for the opercular movements to be counted for at least half a minute, but in some cases it was found necessary to hold the fish gently in a blunt pair of forceps while making the count' (Jones, 1938). Nowadays, the sensitivity of fish to such treatment is well appreciated and clearly the results obtained with such techniques would be affected. Furthermore, the irregular breathing of sticklebacks and other fish also make such counts very difficult.

The use of a number of physiological techniques now renders such methods of observation unnecessary. Some of these have been reviewed by Heath (1972) and amongst them he includes a method developed for recording pressure changes of the buccal cavity (Hughes & Shelton, 1958). A study using a simple electrical monitoring system for ventilatory frequency is described by Morgan & Kühn (1974). This method gave clear results when testing responses of fish to heavy metal pollutants over several days (Fig. 6). However, an important aspect of this problem is the realisation that frequency alone is only one of the parameters involved. Equally important is the

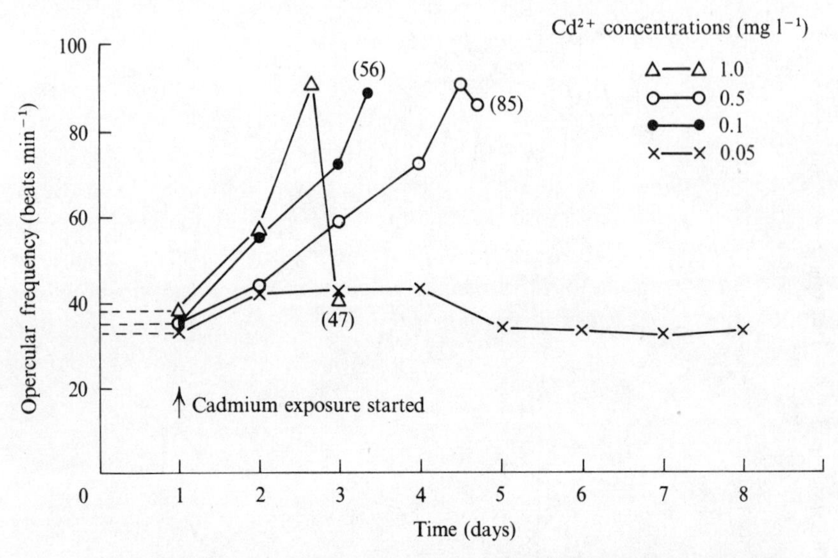

Fig. 6. Daily average opercular frequencies of *Micropterus salmoides* exposed to different solutions of cadmium. Numbers in brackets denote times of death for individual fish. (From Morgan & Kühn, 1974.)

ventilatory stroke volume; the product of these two gives the ventilation volume which is a very relevant parameter because it indicates the volume of polluted water passing across the gills during a given time.

Methods of measuring ventilation volume have been divided into direct and indirect methods (Hughes & Knights, 1968; Hughes, 1970). The former are probably more accurate (Davis & Cameron, 1970) but they tend to cause greater interference with the normal functioning of the ventilatory mechanism. The most commonly used indirect method (Saunders, 1961) depends on simultaneous measurement of the oxygen consumption (V_{O_2}) and the oxygen content of the inspired (C_{insp}) and expired water (C_{exp}). Hence:

$$\text{ventilation volume} = V_{O_2}/(C_{insp} - C_{exp}).$$

The measurement of oxygen consumption of fish is fraught with many difficulties (Fry, 1957). Of the methods available, that using a continuous flow respirometer, similar to that shown in Fig. 2, is probably the most reliable for the measurement of resting or routine metabolism. Ideally measurements should be made at different levels of activity, but more complex apparatus is required for such studies (Brett, 1964; Farmer & Beamish, 1969).

Cardiac function during pollution

On the blood side of the gas exchange system, the easiest recording to make is the electrocardiogram which provides information about heart rate. Recordings of this kind have shown that during pollution treatment there is a bradycardia similar to that which commonly occurs during hypoxia (Skidmore, 1970). Evidence can also be obtained from such physiological methods when the cardio-ventilatory system shows some malfunction. The presence of bradycardia during cadmium poisoning of tautog in spite of no apparent histopathological changes in the gills (unpublished data, quoted by Eisler, 1971) suggests some tissue hypoxia. Such methods are also being employed in studies of pollution by French fish physiologists (Labat, Dazarola & Chatelet, 1973).

Recently we have analysed the relationship between the rhythms of ventilation and heart beat using another parameter which is of interest in this context – the cardio-ventilatory coupling which increases during hypoxia in trout (Hughes, 1972*b*, 1973). An increased relationship was noted by Skidmore (1970) who drew attention to the occurrence of synchrony (i.e. 100 % coupling) of the two rhythms (Fig. 7*a*) but at that stage no method was available for expressing this relationship quantitatively.

We have been able to repeat his result in some recent studies (Hughes & Adeney, unpublished data) and have confirmed that when trout were subjected to such concentrations of zinc there was a bradycardia accompanied by an increase in coupling (Fig. 7*b*). The heart never became completely synchronised with the ventilatory rhythm. This technique makes it possible to analyse data throughout long experiments, but whether it is sufficiently sensitive to detect the effects of lower concentrations of pollutants has not yet been established.

Cannulation of the blood system

A more sensitive indicator of cardiac function is measurement of oxygen levels within the blood, and the concentration of different chemicals including pollutants and metabolites. Such a technique enabled Skidmore to confirm that the oxygen level in the blood leaving the gills was reduced as a result of zinc pollution; it also enabled him to inject zinc into the circulation and so demonstrate that it did not have the serious consequences that were found when such concentrations were bathing the outer membranes of the gills. He thus concluded that zinc reduced the effectiveness of the gills in the transfer of oxygen. The view that fish die in such waters as a result of hypoxia has also been supported by Burton, Jones & Cairns (1972). Heath & Hughes (1973)

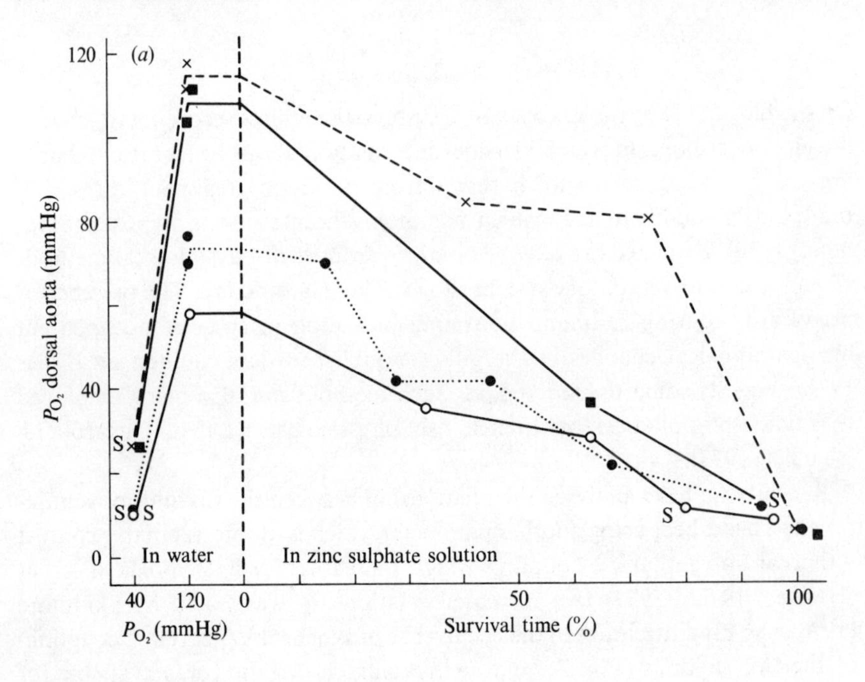

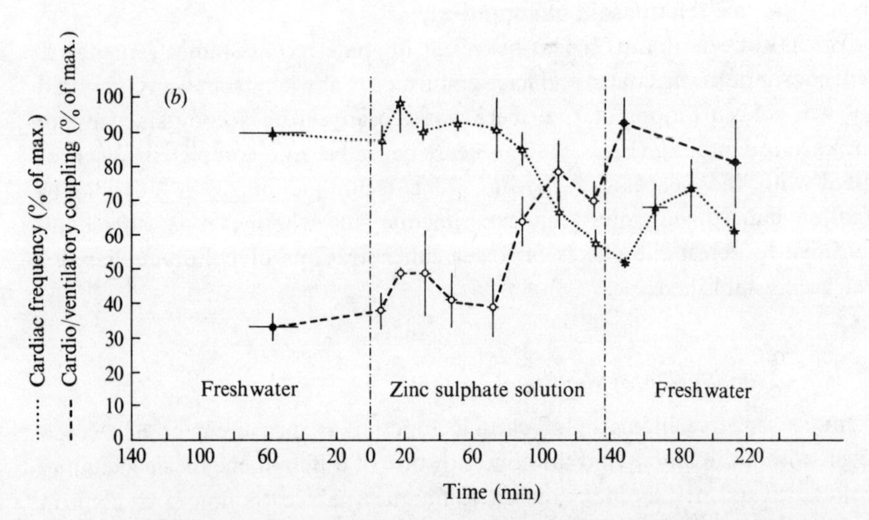

Fig. 7. (*a*) Oxygen tension (P_{O_2}) in dorsal aortic blood of four rainbow trout in zinc-free water and zinc sulphate solution (40 ppm) until immobilisation of the opercula at 100 % survival time. S, synchrony. (From Skidmore, 1970.) (*b*) Plot to show the effect of zinc sulphate (40 ppm) on the cardiac frequency and percentage coupling between the cardiac and ventilatory rhythms. Values are plotted as percentages of the maximum cardiac frequency and of the maximum percentage coupling observed during each experiment. Values ± 1 standard error are shown in each case. Based upon results from five fish. (From Hughes, 1976.)

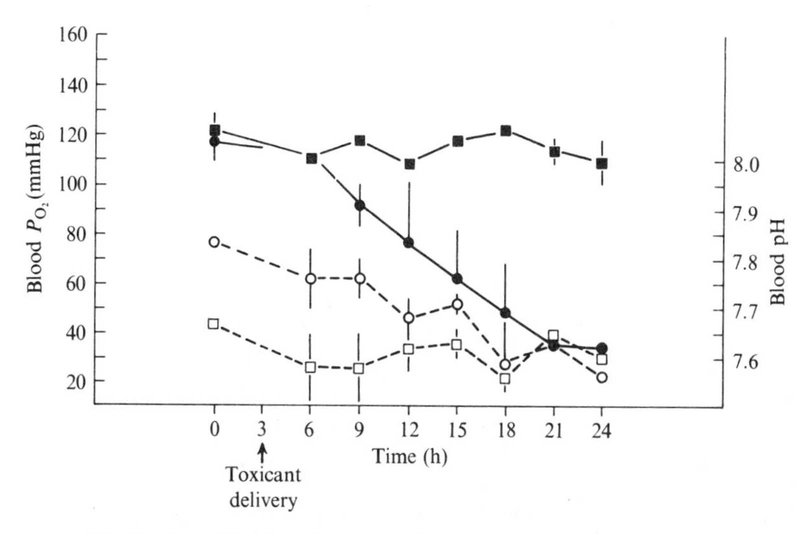

Fig. 8. Plots showing changes in blood P_{O_2} (solid symbols) and pH (open symbols) as a result of treatment of rainbow trout with 1.43 ppm zinc. Squares represent control values, circles experimental values. (From Sellers, Heath & Bass, 1975.)

concluded as a result of similar studies using blood samples from the dorsal aorta that cardiovascular adjustments may be a limiting factor in trout subjected to thermo-pollution. More recently using lower concentrations (1.43 ppm) of zinc, Sellers, Heath & Bass (1975) have shown (Fig. 8) that in addition to changes in blood oxygen tension (P_{O_2}) there is also a marked decrease in blood pH following treatment with polluted water. The decrease in pH is probably related to increases in concentration of metabolic products such as lactic acid. Such changes in blood lactate are known for coho salmon when the fish is in hypoxic environments such as it might encounter in rivers or estuaries (Fig. 9). Smith *et al.* (1973) emphasised the value of such studies as background information for the use of salmon as bioassay organisms for multifactor studies on sublethal pollution. They pointed out that no single physiological criterion is sufficient to describe all of the problems faced by the salmon even when they are confronted with single factor pollution such as low dissolved oxygen. Extensive studies have been carried out on the blood of many types of fish and have shown that some pollutants can affect the haematocrit, haemoglobin concentration and the red and white cell count (e.g. Larsson, this volume; Mitrovic, 1972; C. L. Mahajan, unpublished observations on Indian air-breathing fishes). Such measurements can become part of a routine analysis of fish from waters suspected of pollution. Samples are best taken from indwelling cannulae but this is not always possible.

In fact, sampling of blood from cannulated fish is not easy and there have

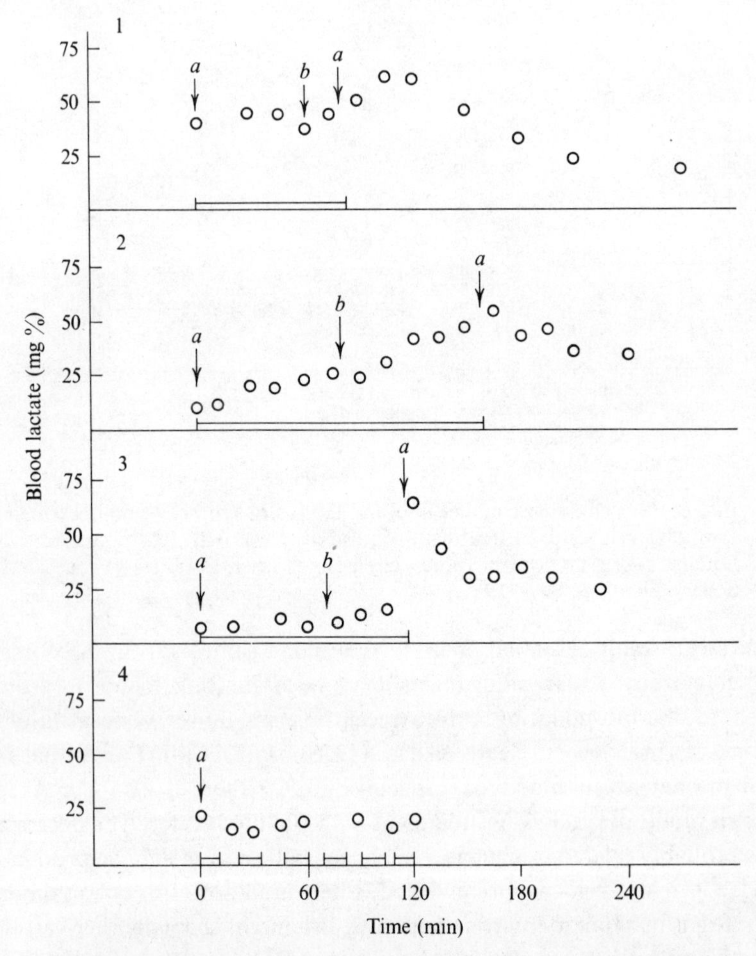

Fig. 9. Changes in blood lactate levels in four individual coho salmon (*Oncorhynchus kisutch*) swimming for various periods (indicated by bars) in: (*a*), elevated (65 % saturation) and (*b*), low (50 % saturation) dissolved oxygen levels. Temperature range, 12–14 °C; salinity, 20–28 ‰. (Reproduced from Smith *et al.*, 1973 by kind permission of the FAO.)

been criticisms of the results of chronic experiments with trout. But *Torpedo* and *Raia* with cannulae implanted in the dorsal aorta remain in excellent condition for several weeks. New techniques are also being developed for trout which promise to give better results during long-term experiments (Soivio, Nyholm & Westman, 1975). Studies of respiratory properties of the blood are little affected by this problem and indicate that the repetition of standard determinations such as P_{50} and the Bohr effect can show significant differences between fish kept in clean and polluted waters for long periods.

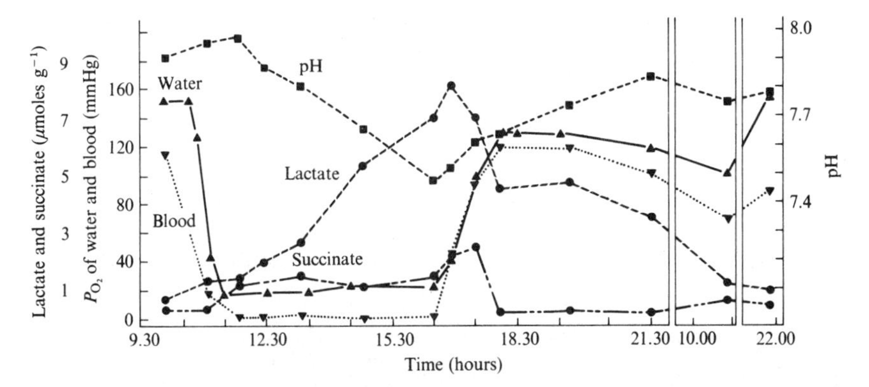

Fig. 10. *Torpedo marmorata*. Plots showing changes in blood parameters during a long hypoxia experiment in which the water P_{O_2} remained below 20 mmHg for about 4 hours. (From Hughes & Johnston, 1976.)

Techniques for making such measurements rapidly and on small volumes of blood are now available (Hughes, Palacios & Palomeque, 1975).

It is evident that the investigation of adaptations of fish to hypoxia from a respiratory point of view is related to the changes which must occur in polluted environments. As Smith *et al.* (1973) also emphasised, studies of this kind are likely to produce relevant information in spite of their main goal being quite different from that of workers more directly concerned with pollution.

A very recent example of this kind has arisen during studies of adaptation of the electric ray (*Torpedo marmorata*) to conditions of low oxygen. We have found that this fish is capable of surviving environments of less than 10 mmHg P_{O_2} for periods of several hours. In addition to the obvious compensatory response of the ventilatory and cardiac pumps (Hughes, 1973) we have now been investigating changes in the blood as indicated by samples drawn from the dorsal aorta. Fig. 10 shows how the dorsal aorta oxygen tension closely follows that of the environmental P_{O_2}. After an initial increase, probably resulting from hyperventilation, it shows a marked decrease similar to that observed in polluted trout by Sellers *et al.* (1975). In addition we measured (Hughes & Johnston, 1976) changes in lactate and succinate levels during this hypoxia experiment. As is clear from Fig. 10, the lowering in blood pH reflects the gradual increase in blood lactate and is indicative of an increase in the anaerobic metabolism of the fish. A new finding in these studies was that there is also an increase in blood succinate which presumably indicates the presence of alternative metabolic pathways for energy release under conditions of low oxygen. *Torpedo* proved to be an excellent fish for making such investigations and possibly provides a model of the sort of responses which may occur in other fish.

This last example is a very recent case which justifies the general approach that I have tried to emphasise in this paper, namely that investigations of respiratory problems for their own sake will produce information that can have relevance to the understanding of the effects of pollution on living organisms.

Whilst it is inevitable that a given group of workers will pay especial attention to a given aspect of fish physiology with which they are concerned, in our case the respiratory system, it is important that they should be broad-minded enough to appreciate other possibilities. A recent example of this kind arose in connection with some studies made at Bristol on the effects of ammonia pollution on rainbow trout. Although this pollutant has chronic effects on the gill secondary lamellae (Smart, 1976), mainly the contractile and skeletal systems of the pillar cell system, it seems probable that the acute lethal action of ammonia is due to changes in brain metabolism. It was found that there was a depletion of energy-rich phosphates in the brains of fish exposed to acutely lethal concentrations. This example also illustrates the way in which the study of respiration must look at all levels of the respiratory chain from the gill membranes to the cellular mechanisms of particular organ systems (Hughes, 1964, 1973).

Gill morphometry

Although in the strict sense not a physiological technique, the study of the morphometrics of respiratory surfaces in relation to their role in gas exchange has proved invaluable to both morphologists and physiologists (Hughes, 1972*a*; Weibel, 1973). In the case of fish gills it can provide information about: (*a*) the surface area across which gas exchange can occur between the water and blood, and (*b*) the diffusion distance for oxygen and carbon dioxide between these two media. A term used by respiratory physiologists in this connection is the oxygen diffusing capacity (D_{O_2}) which is defined by the relationship $D_{O_2} = \dot{V}_{O_2}/\Delta P_{O_2}$, i.e. the ratio between the amount of oxygen transferred to the difference in oxygen tension between the water and blood. From a morphometric point of view the pathway of gas transferred includes the water, the tissue barrier, the blood plasma and finally the red blood cells. Measurements can be made of these distances and together with the area (A) and appropriate diffusion constant (K), they enable morphometric estimates of the diffusing capacity to be made (Hughes, 1972*a*; Hughes & Perry, 1975) using the relationship: $D_{O_2} = K \cdot A/t$, where t is the diffusion distance.

In nearly all cases these values have been higher than the corresponding values obtained by physiological methods. It appears that morphometrically determined values indicate the maximum possible capacity.

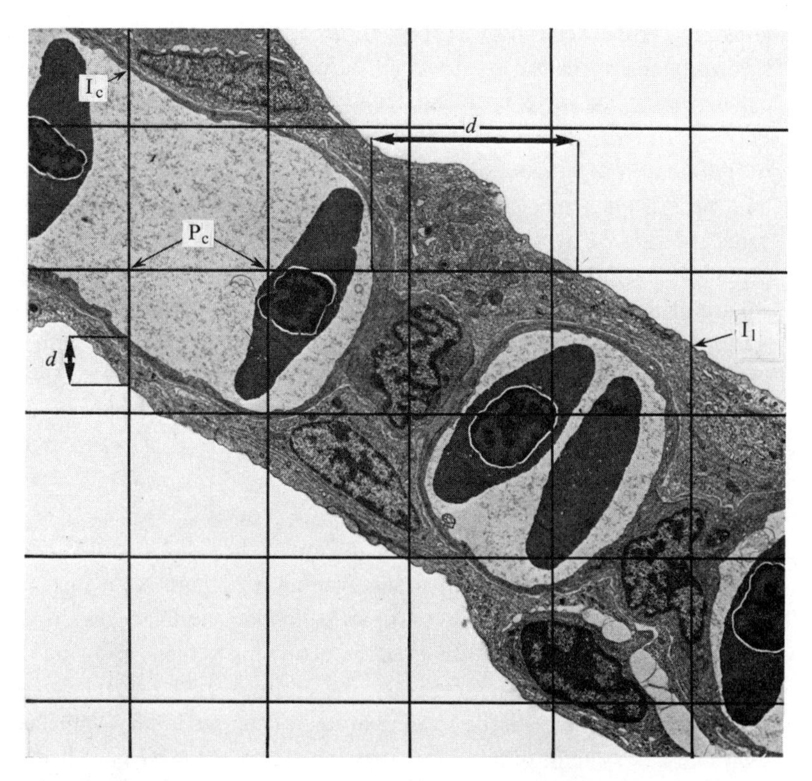

Fig. 11. Electronmicrograph of a tench secondary lamella with superimposed rectilinear grid to show some measurements used in gill morphometry. Thick arrows give two examples of distances (*d*) measured for the determination of harmonic mean thickness of the tissue barrier. Small arrows indicate intersections with the secondary lamellar surface (I_l) and of a blood channel surface (I_c). Two points which fall within the blood spaces (P_c) are also indicated. (From Hughes, 1972*a*.)

In the case of polluted gills, damage to the surface of the secondary lamellae has been recognised by many investigators (Brown, Mitrovic & Stark, 1968; Skidmore & Tovell, 1972), who emphasised separation of the surface epithelium and the oedematous condition of the gill following exposure to relatively high concentrations of heavy metal and some other pollutants. Although such marked changes are clearly visible, the quantitative description of the degree of damage and the recognition of slight damage is difficult; it would also be valuable to know the particular way in which these changes influence the secondary lamellar structure. It is clear, for example, that the diffusing capacity will be reduced and consequently the fish subjected to some form of hypoxia (Hughes, 1973).

It is only during the last few years that the morphometric investigation of gill structure has been started (Hughes, 1966, 1972*a*). Initially this involved the measurement of the harmonic mean diffusion distance but recently far more extensive studies have been carried out (Hughes & Perry, 1976; Hughes, Perry & Brown, 1976). The essential technique involves the use of sections of gill material which are then viewed on a projection microscope and analysed using a special form of grid. Most recently we have used the so-called Merz grid because in this case the sections were made with a specific orientation of the gill material. Where the orientation is more random a square grid (Fig. 11) is more appropriate as in the previous work (Hughes, 1972*a*). For each section, counts are made of intersections with different surfaces and of points where they fall on specific items in the section. In this way comparisons are possible between material from normal fish and those treated with pollutant. In order to illustrate the sort of result that can be obtained, I will give three specific examples:

(*a*) *Point counting* has shown, for example, that the so-called oedema involves an absolute increase in tissue volume, i.e. there is an increase in the relative volume occupied by tissue within the lamellar region (V_{tis}/V_{LR}) following nickel treatment – from 0.26 in controls to 0.53 following treatment in 3.2 ppm nickel. At zero recovery the controls had a V_{tis}/V_{OPS} ratio (i.e. the relative volume of tissue to the volume of the secondary lamellae outside the pillar cell system) of 0.817 whereas those treated in 3.2 ppm had values of 0.715, i.e. there was only about 10 % more non-tissue constituent following this treatment. Thus a quantitative estimate can be given of what happens during this stage of pollutant action. In fact it shows that there is not a true oedema, which generally suggests a much greater relative increase in the non-tissue constituent.

(*b*) *Intersections* of the grid lines with the external surface were taken as points from which measurements of the minimum distance to the nearest red blood cell were made in order to calculate the harmonic mean thickness Intersections with the outer surface also give estimates of the relative size of lamellar surfaces; absolute values for surface area are not required for comparison between different treatments. We have introduced the term D_{rel} to indicate the ratio between diffusing capacity of the control, i.e. the normal gill, and that of pollutant-treated fish. In Fig. 12 recovery following nickel pollution is clearly visible as D_{rel} returns from a very low value of 0.124 to that of the control value (1.0) or even greater.

(*c*) As mentioned earlier, the availability of techniques sufficiently sensitive to detect changes brought about by low concentrations of pollutants are now required. We were fortunate to obtain cadmium-treated fish from Stevenage following treatment for 7 months and found that D_{rel} after no recovery

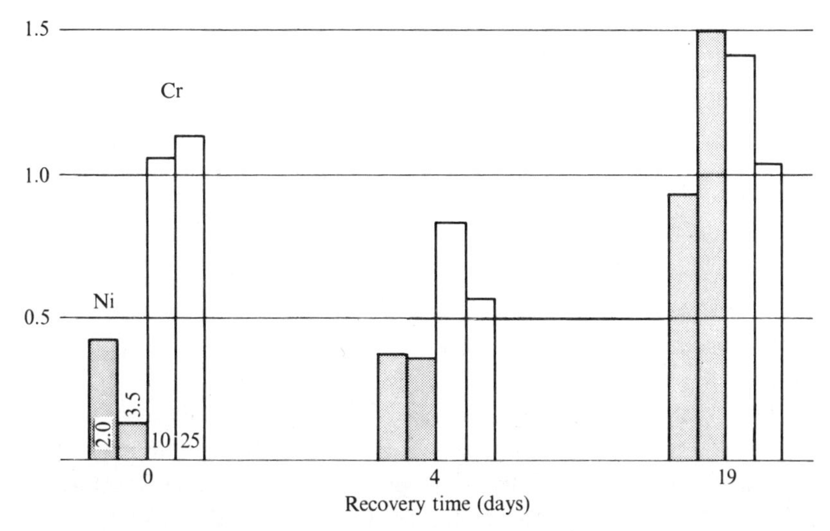

Fig. 12. Histograms to show changes in relative diffusing capacity (D_{rel}) of rainbow trout following treatment for 5000 min with the concentrations of nickel and chromium (mg l^{-1}) indicated in the columns. Recovery following the marked reduction in diffusion capacity after nickel contrasts with the relatively slight changes following chromium pollution.

from a solution containing 0.002 mg calcium l^{-1} was 1.359 with a standard deviation of 0.178. Consequently, a fish in this solution has a diffusing capacity which is about one-third greater than controls. This somewhat surprising result suggests that fish kept in this very low concentration are in fact in a more favourable environment from a respiratory point of view than those fish kept in normal Stevenage water.

These three examples have been chosen from a large amount of data that has been compiled in such studies and at the moment we do not have a full understanding of the implications of this type of analysis. Such studies are extremely laborious and require very patient and painstaking observations and consequently the number of fish that have been investigated is relatively small as yet, although totalling forty-five specimens of rainbow trout. Further consolidation of the results and the finer analysis of the data require a study of larger numbers. In the future we also hope to concentrate efforts on those gill parameters which seem to be most sensitive to these pollutants.

A further possibility is to analyse simultaneously the physiological properties of the gills of fish from such environments. Such a physiological/morphological approach would probably be more profitable as it would provide not only information about the morphometrics but also the functioning of the gill system. Because of the homeostatic mechanisms involved (Hughes, 1964) the fish respiratory system is remarkably flexible and capable of modification in relation to environmental changes. Morphometric investigations are of necessity static and unable to provide data about the dynamic condition of the normal fish, but a combination of the two approaches would provide information which is greater than the sum of the two studies carried out separately. Such approaches could clearly be applied to other organ systems using similar morphometric methods and might well produce even more striking results indicative of the effects of sublethal doses of pollutants.

We have seen in these morphometric studies the importance of the water/blood barrier. Another portion of the pathway of oxygen from the water to combination with haemoglobin is the reaction within the red blood cells and some authors have considered this to be the limiting part of the whole transfer process. The investigation of the velocity of this reaction in fish has only recently been carried out (Hughes & Koyama, 1975). This study showed how isolated secondary lamellae were ideal for the use of a microphotometric technique developed by Mochizuki, Tazawa & Ono (1974). One surprising result of the further analysis of these experiments (Hughes & Hills, unpublished observations) is that this preparation can provide important information about the properties of the barrier itself and consequently could be very valuable for comparing the physiological properties of secondary lamellae in fish from normal and polluted waters.

Conclusion

This account has inevitably been influenced by my own particular bias but I hope that some of the examples given indicate the value of respiratory studies to the solution of some pollution problems. Furthermore, I hope that it justifies some faith in a system of 'free' research and that such a system can produce results of relevance to more applied problems. In this context we would do well to remember the words of Sir Vincent Wigglesworth (1971) at the end of the first George Bidder Lecture, 'I hold that there are still unexpected discoveries to be made, and that there is still room for the enquiring mind and the untrammelled researches of the experimental biologist'. However, it is equally clear that the overall research effort must not be made in this direction alone and to the exclusion of direct applied researches. In the Universities I would suggest that there should be encouragement of the more academic research which adds to the total fund of knowledge rather than that such scientists be over-distracted from their primary research objectives, as this would lead to an erosion of the capital of knowledge which has accumulated over the years as a result of this basic type of research. On the other hand sufficient facilities and appreciation should be given to those researches of a more directly applied nature as this has sometimes been neglected in the past. In other words I suggest that both types of research be encouraged and most importantly that arrangements be made for workers of all types to meet and discuss their research interests.

In summary then, I suggest interchange, interaction and integration of these different approaches. In this way I believe that all the available man-power would be most effectively harnessed and would lead to the most rapid evolution of methods and approaches not only for the solution of practical problems but also for making basic discoveries. It would be a more stable situation than one which oscillates between emphasis on the two extremes according to the prevalent politics.

Much of the work discussed here was supported by funds from the Natural Environment Research Council. I should like to take this opportunity of thanking not only the Council but also those research assistants and technicians who have been able to help in this work.

References

Bert, P. (1871*a*). Sur les phénomènes et les causes de la mort des animaux d'eau douce que l'on plonge dans l'eau de mer. 1st Note. *Comptes Rendus Hebdomadaire des Séances de l'Académie des Sciences*, **73**, 382–5.

Bert, P. (1871*b*). Sur les phénomènes et les causes de la mort des animaux d'eau douce que l'on plonge dans l'eau de mer. 2nd Note. *Comptes Rendus Hebdomadaire des Séances de l'Académie des Sciences*, **73**, 464–7.

Brett, J. R. (1964). The respiratory metabolism and swimming performance of young sockeye salmon. *Journal of the Fisheries Research Board of Canada*, **21**, 1183–1226.

Brown, V. M., Mitrovic, V. V. & Stark, G. T. C. (1968). Effects of chronic exposure to zinc on toxicity of a mixture of detergent and zinc. *Water Research*, **2**, 255–63.

Burton, D. T., Jones, A. H. & Cairns, J. Jr (1972). Acute zinc toxicity to rainbow trout (*Salmo gairdneri*): confirmation of the hypothesis that death is related to tissue hypoxia. *Journal of the Fisheries Research Board of Canada*, **29**, 1463–6.

van Dam, L. (1938). On the utilization of oxygen and regulation of breathing in some aquatic animals. Dissertation, Gröningen.

Davis, J. C. & Cameron, J. N. (1970). Water flow and gas exchange at the gills of rainbow trout, *Salmo gairdneri*. *Journal of Experimental Biology*, **54**, 1–18.

Eisler, R. (1971). Cadmium poisoning in *Fundulus heteroclitus* (Pisces: Cyprinodontidae) and other marine organisms. *Journal of the Fisheries Research Board of Canada*, **28**, 1225–34.

Farmer, G. J. & Beamish, F. W. H. (1969). Oxygen consumption of *Tilapia nilotica* in relation to swimming speed and salinity. *Journal of the Fisheries Research Board of Canada*, **26**, 2807–21.

Fry, F. E. J. (1957). The aquatic respiration of fish. In *Physiology of fishes* (ed. M. E. Brown), vol. 1, pp. 1–63. Academic Press, New York & London.

Heath, A. G. (1972). A critical comparison of methods for measuring fish respiratory movements. *Water Research*, **6**, 1–7.

Heath, A. G. & Hughes, G. M. (1973). Cardiovascular and respiratory changes during heat stress in rainbow trout (*Salmo gairdneri*). *Journal of Experimental Biology*, **59**, 323–38.

Herbert, D. W. M. & Merkens, J. C. (1961). The effect of suspended mineral solids on the survival of trout. *International Journal of Air and Water Pollution*, **5**, 46–55.

Hughes, G. M. (1964). Fish respiratory homeostasis. *Symposia of the Society for Experimental Biology*, **18**, 81–107.

Hughes, G. M. (1966). The dimensions of fish gills in relation to their function. *Journal of Experimental Biology*, **45**, 317–33.

Hughes, G. M. (1970). A comparative approach to fish respiration. *Experientia*, **26**, 113–22.

Hughes, G. M. (1972*a*). Morphometrics of fish gills. *Respiration Physiology*, **14**, 1–25.

Hughes, G. M. (1972*b*). The relationship between cardiac and respiratory rhythms in the dogfish *Scyliorhinus canicula* L. *Journal of Experimental Biology*, **57**, 415–34.

Hughes, G. M. (1973). Respiratory responses to hypoxia in fish. *American Zoologist*, **13**, 475–89.

Hughes, G. M. (1975). Coughing in the rainbow trout (*Salmo gairdneri*) and the influence of pollutants. *Revue Suisse de Zoologie*, **82**, 47–64.

Hughes, G. M. (1976). Fish respiratory physiology. In *Proceedings of 50th Anniversary Meeting of the Society for Experimental Biology, Cambridge, 1974* pp. 235–45. Pergamon Press, Oxford.

Hughes, G. M. & Johnston, I. A. (1976). Some effects of hypoxia on the blood chemistry of the electric ray (*Torpedo marmorata*). (In preparation.)

Hughes, G. M. & Knights, B. (1968). The effect of loading the respiratory pumps on the oxygen consumption of *Callionymus lyra*. *Journal of Experimental Biology*, **49**, 603–15.

Hughes, G. M. & Koyama, T. (1975). Gas exchange of single red blood cells within secondary lamellae of fish gills. *Journal of Physiology, London*, **246**, 82–83P.

Hughes, G. M., Palacios, L. & Palomeque, J. (1975). A comparison of some methods for determining oxygen dissociation curves of fish blood. *Revista Espanola de Fisiologia*, **31**, 59–65.

Hughes, G. M. & Perry, S. F. (1975). Morphometric study of the effects of certain heavy metal pollutants on trout gills: model and methods. *Journal of Experimental Biology*, **64**, in press.

Hughes, G. M., Perry, S. F. & Brown, V. M. (1976). A morphometric study of effects of nickel, chromium and cadmium on the secondary lamellae of rainbow trout gills. *Water Research*, in press.

Hughes, G. M. & Shelton, G. (1958). The mechanism of gill ventilation in three freshwater teleosts. *Journal of Experimental Biology*, **35**, 807–23.

Jones, J. R. E. (1938). The relative toxicity of salts of lead, zinc and copper to the stickleback (*Gasterosteus aculeatus* L.) and the effect of calcium on the toxicity of lead and zinc salts. *Journal of Experimental Biology*, **15**, 394–407.

Jones, J. R. E. (1939). The relation between the electrolytic solution pressures of the metals and their toxicity to the stickleback (*Gasterosteus aculeatus* L.). *Journal of Experimental Biology*, **16**, 425–37.

Jones, J. R. E. (1964). *Fish and river pollution*. Butterworth, London.

Labat, R., Dazarola, G. & Chatelet, A. (1973). Principe d'un dispositif technique servant a l'étude des effluents toxiques dans la pollution de l'eau. *Bulletin de la Société d'Histoire Naturelle de Toulouse*, **109**, 1–2.

Lloyd, R. (1961). Effects of dissolved oxygen concentrations on the toxicity of several poisons to rainbow trout (*Salmo gairdneri* Richardson). *Journal of Experimental Biology*, **38**, 447–55.

Mitrovic, V. V. (1972). Sublethal effects of pollutants on fish. In *Marine pollution and sea life* (ed. M. Ruivo), pp. 252–7. Fishing News (Books) Ltd, London.

Mochizuki, M., Tazawa, H. & Ono, T. (1974). Microphotometry for determining the reaction rate of O_2 and CO with red blood cells in the chorioallantoic capillary. In *Oxygen transport to tissue* (ed. D. F. Bruley & H. I. Bicher), pp. 997–1006. Plenum Press, London.

Morgan, W. S. G. & Kühn, P. C. (1974). A method to monitor the effects of toxicants upon breathing rate of largemouth bass (*Micropterus salmoides*, Lacépède). *Water Research*, **8**, 67–77.

Saunders, R. L. (1961). The irrigation of the gills in fishes. I. Studies of the mechanism of branchial irrigation. *Canadian Journal of Zoology*, **39**, 637–53.

Sellers, C. M., Heath, A. G. & Bass, M. L. (1975). The effect of sublethal concentrations of copper and zinc on ventilatory activity, blood O_2 and pH in rainbow trout (*Salmo gairdneri*). *Water Research*, **9**, 401–8.

Skidmore, J. F. (1970). Respiration and osmoregulation in rainbow trout with gills damaged by zinc sulphate. *Journal of Experimental Biology*, **52**, 484–94.

Skidmore, J. E. & Tovell, P. W. A. (1972). Toxic effects of zinc sulphate on the gills of rainbow trout. *Water Research*, **6**, 217–30.

Smart, G. R. (1976). The effect of ammonia exposure on gill structure of the rainbow trout (*Salmo gairdneri*). *Journal of Fish Biology*, in press.

Smith, L. S., Cardwell, R. D., Mearns, A. J., Newcomb, T. S. & Watters, K. W. Jr (1973). Physiological changes experienced by Pacific salmon migrating through a polluted urban estuary. In *Marine pollution and sea life*, (ed. M. Ruivo), pp. 322–5. Fishing News (Books) Ltd, London.

Soivio, A., Nyholm, K. & Westman, K. (1975). A technique for repeated sampling of the blood of individual resting fish. *Journal of Experimental Biology*, **63**, 207–17.

Warren, C. E. (1971). *Biology and water pollution control*. W. B. Saunders, Philadelphia.

Weibel, E. R. (1973). Morphological basis of alveolar–capillary gas exchange. *Physiological Reviews*, **53**, 419–95.

Wigglesworth, V. B. (1971). Experimental biology pure and applied. *Journal of Experimental Biology*, **55**, 1–12.

INDEX